VHF Afloat
Sara Hopkinson

3rd Edition

VHF Afloat

Sara Hopkinson

WILEY ✸ NAUTICAL

Published by John Wiley & Sons Ltd
The Atrium, Southern Gate, Chichester, West Sussex PO19 8SQ, England
Telephone (+44) 1243 779777
Email (for orders and customer service enquiries): cs-books@wiley.co.uk
Visit our Home Page on www.wiley.com

We are also grateful to Northshore Yachts for use of their 3D model of the Southerly 38 swing keel cruising yacht in the illustrations.

Other Wiley Editorial Offices
John Wiley & Sons Inc., 111 River Street, Hoboken, NJ 07030, USA
Jossey-Bass, 989 Market Street, San Francisco, CA 94103-1741, USA
Wiley-VCH Verlag GmbH, Boschstr. 12, D-69469 Weinheim, Germany
John Wiley & Sons Australia Ltd, 165 Cremorne Street, Richmond Vic. 3121, Australia
John Wiley & Sons (Asia) Pte Ltd, 2 Clementi Loop #02-01, Jin Xing Distripark, Singapore 129809
John Wiley & Sons Canada Ltd, 22 Worcester Road, Etobicoke, Ontario, Canada, M9W 1LI

ISBN: 978-0-470-75858-8

Typeset and design by PPL, illustrations by Greg Filip/PPL
Printed and bound in Singapore by Markono Print Media Pte Ltd
Typeset in Humnst777EU
This book is printed on acid-free paper responsibly manufactured from sustainable forestry in which at least two trees are planted for each one used for paper production.

Acknowledgements
The author would like to thank Gwen Caddock, Robin Cole of Precision Navigation Ltd, and Jonathon Dyke of Suffolk Yacht Harbour. The International VHF Frequencies are reproduced courtesy of the MCA (Maritime and Coastguard Agency) as is the map of the GMDSS Sea Areas.

Photos supplied by Icom UK Ltd.

You can experience unlimited hands-on practice of DSC operations using a Simulator program running on your PC. LightMaster Software is the UK's leading producer of training software for marine leisure. LightMaster have worked very closely with the RYA and with the MCA to produce training and simulation software which has been adopted as the standard for use in VHF and Radar courses. Other products comprise animated training programs covering Collision Regulations, Buoys, and GPS Navigation. Single-user versions of each program can be purchased at low cost to provide valuable preparation for courses and realistic ongoing revision, without the risks such as false alerts or collisions which might result from practising on real equipment or vessels. All LightMaster software runs on standard PCs with Windows™ and is available from: **www.lightmaster.co.uk**

18 Stanley Gardens, South Croydon, Surrey, CR2 9AH, Tel: 020 8405 8200, Fax: 020 8405 8300, sales@lightmaster.co.uk

Contents

Foreword VHF is Wonderful ...VHF-DSC is better!... 6

1   Licences and certificates 8

2   Types of VHF set 10

3   What is the range of the set? 12

4   How to begin using a VHF radiotelephone 14

5   The DSC Controller 18

6   Which channel do I use? 20

7   What do I say? ...Prowords 24

8   Ship-to-Ship routine communications 26

9   Routine communications with HM Coastguard 29

10 Routine communications with marinas, ports and harbours 32

11 On passage withyacht Sierra 34

12 Distress proceduresand Coastguardacknowledgement 42

13 What do I do ifI hear a Mayday? 45

14 The Mayday Relay...'Urgency Alert' 47

15 Pan Pan... 'Urgency Alert' 49

16 Securite ...'Safety Alert' 52

17 SIERRA ...the return! An eventful homeward passage 54

18 EPIRBs.. 60

19 SARTs.. 62

20 NAVTEX 64

21 Radio channels 66

22 Glossary.. 68

     Questions 70

     Answers... 74

     Useful Addresses 76

     Mayday Procedure Card 77

Foreword
VHF is Wonderful
...VHF-DSC is better!

The VHF radiotelephone has been in common use on small boats for many years. Its use is essentially simple and straightforward but the new user must learn a few rules and procedures, pass a practical assessment and a simple test, use common sense and avoid chatting.

All very easy!

VHF is invaluable in the unlikely event of a major life-threatening disaster when a **Mayday** can be sent to gain the assistance of the rescue services and vessels in the vicinity. It is also useful for routine communications with marinas, Coastguards, harbour authorities and other boats.

It is an 'open' system of communication so can receive information such as weather forecasts, shipping movements, gale warnings and small craft safety information broadcasts. All skippers in a particular area can be kept informed by these broadcasts. In this way it is of value every day.

This very 'openness' of communication should be remembered by the user. All conversations on a channel can be heard by all those tuned to that channel who are within range. So never say anything that you are not prepared for everyone to hear!

The system is operated on approximately 60 channels, with the radio frequencies pre-tuned into the set. Channels are allocated under international agreement for different uses. It is essential to know these allocations and the rules and procedures for both routine and emergency radio traffic. Failure to follow these rules will result in interference with routine communications and possibly dangerous confusion in an emergency situation.

Each of these channels can handle only one conversation at a time – so while you are occupying a channel no-one else can use it. Hence the rule not to chat. Pass the essential information and then leave the channel free for someone else.

Over the years there have been many minor changes and developments in both procedures and equipment. These have been designed to ease the overloading of the airwaves and to keep pace with changes in technology.

These developments include satellite communications, e-mail and the ever-present mobile phone. I have sailed on yachts with as many mobile phones as crew! Love them or hate them, they can be invaluable for keeping in contact with those at home, but a mobile phone is a very poor substitute for a VHF in a distress situation. Some people say 'they are better than nothing' which is obviously true but they still offer very poor communications in emergency situations. Coastguard stations can be contacted, while the battery holds out, but lifeboats and helicopters do not use mobile phones, and accurate direction finding on the signal is impossible.

The system has seen major changes following the introduction of **GMDSS** (the Global Maritime Distress and Safety System) in February 1999. This world-wide system was devised by the International Maritime Organisation (IMO) in conjunction with the International Telecommunication Union (ITU). The purpose of GMDSS is to ensure that a vessel can initiate a distress alert automatically by at least two independent means to a Rescue Co-ordination Centre and that all vessels in the vicinity will be aware of the situation.

This system is compulsory for ships over 300 GRT and vessels which carry more than 13 passengers. To achieve these broad aims, GMDSS uses various types of communication and information systems:

- **Radio equipment** of a suitable type to broadcast over the range required – VHF or MF – both fitted with digital selective calling (DSC). This initiates the call more quickly, identifies the calling station and includes the position of the vessel in a distress call if a GPS is connected.

- **Distress alerting equipment** – EPIRB, in addition to the radio.

- **Search location equipment** – SART, in addition to the EPIRB.

- **Safety information system** – Navtex.

An outline knowledge of these types of equipment is important for all skippers and is essential for those wanting to upgrade their existing VHF licence or to take the Short Range Certificate course. The course content was agreed by CEPT (Conference of European Posts and Telecommunications). In this book, each type of call is illustrated with the Lightmaster DSC Computer Simulation. Whichever set you have on your boat, the facilities will be similar, check the instruction book for specific details.

The GMDSS system is not compulsory for small yachts and motorboats but there are great benefits to be had and increasingly the craft without a DSC radio will be at a disadvantage and severely limited in its ability to communicate, even in distress situations. You will want one of the new sets!

RAINWAT

Rainwat, that is the **R**egional **A**rrangement concerning the Radiotelephone service on **In**land **Wat**erways, is an agreement reached in 2000 between several European countries.

The waterways this particularly concerns are those navigable by ships, such as the Rhine and the Danube, where shipping movements are tightly monitored. Some of this monitoring can be done automatically with the use of an automatic radiotelephone identification system or **ATIS**. This type of radio produces an ID signal as the PTT button is released. The Netherlands is a signatory to this agreement and those on the East Coast have been hearing the tell-tail squawk for a few years now as the Dutch vessels visit in the summer.

At the moment this system is not required for small UK flagged vessels on the Inland waters of Europe. This ID system does not include DSC so should it come into force for visiting yachts and motor cruisers there may be a requirement to have a portable VHF with ATIS to use in the river and canal system in the future.

1 Licences and certificates

For those planning to have a VHF radio on their boat there are some legal requirements covering the equipment, and licences and certificates have to be obtained.

The equipment itself must have a CE mark and Declaration of Conformity, which has replaced the old **UK Type Approval.** This simply means that it must come up to an internationally agreed standard. Sets which do not comply are illegal and may cause interference. If you buy a well-known make from a reputable dealer this will not be a problem. To carry radio equipment on your boat you will require a **Ship Radio Licence** from the Office of Communications (Ofcom). This licence covers all transmitting equipment on the boat so all must be listed on the document. This could include MF/HF long range radios, satellite communications equipment, VHF or VHF-DSC, portable VHF, radar, EPIRB and SART. To get a licence contact Ofcom online via www.ofcom.org.uk and it is **free.** The licence remains valid for ten years but needs updating if the equipment, ownership of the vessel or any other details change. You can do this online too. The process can be done by post, but then it is not free!

It is from this licence that the ship's International **call-sign** comes. It is allocated to the boat and not the owner. So if you buy a secondhand boat with a VHF set which is already licensed the call-sign will stay the same but Ofcom will need to know about the change of ownership. This call-sign is made up of numbers and letters. For example: MCDW7. When said over the radio this would be spelled out using the phonetic alphabet and becomes MIKE CHARLIE DELTA WHISKEY SEVEN.

A VHF-DSC radio must have an individual identi-fication number programmed into it before use   .
This **MMSI** number (Maritime Mobile Service Identity) is allocated by the licensing authority. An MMSI has 9 numbers and on a British vessel will start with '232', '233', '234' or '235', followed by 6 digits.

If the only radio that you own is a hand-held set then it will not be covered by a ship radio licence and requires a Ship Portable Radio Licence.

Once installed a VHF set can only be operated by a holder of the appropriate **Certificate of Competence** or by a crew member under their control. This certificate of competence also conveys to the holder the **authority to operate** the set. Someone under 16 may gain the certificate but will have to wait for the authority to operate, which follows automatically after their sixteenth birthday. All certificate holders are bound, by the agreement that they sign, not to divulge information about conversations that they overhear on the radio.

For a standard VHF set the correct certificate was the **Restricted Certificate of Competence in Radiotelephony (VHF only).** This was replaced in September 2000 by the **Short Range Certificate,** so that all new certificate holders are familiar with the VHF-DSC system and qualified to use a DSC set. For the many thousands of people who already hold a VHF certificate there is a short conversion course for the SRC. It is necessary to do this when buying a VHF-DSC set.

The system of assessments for these certificates is run by the Royal Yachting Association **(RYA),** under the authority of the Maritime and Coastguard Agency **(MCA).** The assessment is not difficult and can be taken following a course at an RYA centre. Details of centres offering these courses can be obtained from the RYA. They are mainly the centres that organise the shorebased navigation classes, and sailing schools. The courses usually last no more than a day or perhaps 3 or 4 evenings. They include both theory and practice using a simulator. Each candidate will be assessed by the instructor, and then recommended to the RYA for the award of the certificate, if they come up to the required standard. Some pre-learning before the course is highly recommended. For those who are familiar with the subject already, perhaps from reading this book, it is possible to take the exam direct. Contact the nearest RYA teaching centre.

2 Types of VHF set

Non-DSC sets

Older non-DSC VHF sets on yachts and
motorboats continue to work and are
still legal to use and certificate holders
do not need to do a conversion course
until they upgrade. A **transportable
set** is an invaluable second radio for
use in an emergency when it can be
taken into a liferaft or used on deck to
communicate in a rescue situation. It
is useful for safety reasons in a tender
when going ashore or in a safety boat
when organising dinghy sailing events.
A portable set is a good buy for
non-boat owners who charter
or who go afloat occasionally,
but they have a limited
range and do require
licensing and certification.
Note that the portable set
covered by a ship's radio
licence can only be used
on the vessel covered
by the licence or by its
tender. It is illegal to use
the portable ashore.

VHF-DSC sets

From 2001 all new fixed
radios sold must be VHF-
DSC or be capable of being
converted to DSC by the

addition of an extra 'black box'. This is called the
DSC Controller. It will provide the **digital selective
calling** (DSC) facility which is the special feature of
the new type of set. What this does is to send, on
channel 70, a burst of digital signals in a code to
'call up' another DSC set. This call can be directed
at an individual, using their MMSI, a group of boats
or 'all stations' in an emergency. Once the link has
been established by the digital 'call', normal voice
transmission will be used. The DSC is essentially a
new method of **establishing communications,** more
reliably than was possible before. The digital signals
are of high radio quality and rapid, the alert taking
just 0.5 seconds. It can be used in both routine and
distress situations.

There are different classes of controller with varying
levels of capability for use in different types of
vessel.

• The **Class D** controller is the one designed for
use with VHF on yachts and motorboats who
make passages within VHF range of the coast.
**Fitting one of these is not compulsory on
private boats.**

Other controllers for VHF-DSC are available
to meet the requirements of ships. These
are **Class A and B Controllers,** which have
enhanced capabilities. For ships **over 300
GRT and passenger vessels that carry more
than 13 passengers,** known as 'convention

ships', fitting new equipment became compulsory in February 1999. This equipment included radio and other rescue communication equipment. This is a requirement under the **Global Maritime Distress and Safety System** (GMDSS) and details depend on the areas of the world in which the ship will be operating. Details of mandatory equipment for convention ships are included in the table for the Al area only. **This book covers the use of VHF-DSC radios and other equipment within an Al area.** An outline knowledge of this other equipment is required for the SRC and is covered later in the book.

GMDSS areas

The equipment that a convention ship has to carry depends on the distance from land and the areas of the world in which the vessel will operate:

Al Area: Within radio range of coast stations using VHF-DSC (30 – 40 miles). Vessel has VHF-DSC (using channel 70).	
EPIRB	(Emergency Position Indicating Radio Beacon)
Navtex	(an information system)
SART	(Search And Rescue Radar Transponder)

A2 Area: Within radio range of a coast station using MF (medium frequency) DSC.

A3 Area: Within the sea area covered by the INMARSAT satellite system. (70°N to 70°S)

A4 Area: The rest of the world using HF (high frequency) DSC.

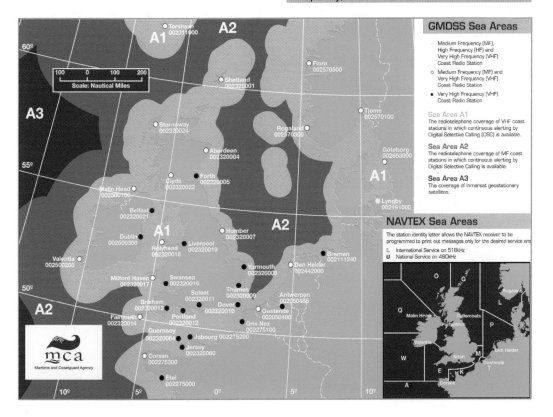

3 What is the range of the set?

Those sailing across an ocean, or even the Bay of Biscay, need radios that transmit over vast distances. These are the A2 or A3 areas and this book does not cover the operation of these long range sets and other communications equipment. The **range** of transmission of the VHF radiotelephone is limited by a number of factors. The **height of the aerial** is very significant as the propagation of the radio waves is only slightly more than 'line of sight'. This includes the aerial height of both the transmitting and the receiving station.

When talking from yacht to yacht expect a range of **15 – 20 miles** with aerials fitted at the tops of the masts. Those commonly fitted to yachts are known as 'unity gain' aerials. They are made of thin wire and sometimes have wind instruments attached. They are recommended because, although the range is not as good as the taller rigid aerials used on motorboats, they cope better with the heeling effect often experienced on yachts! The better range of a 'high gain' motor cruiser aerial is only achieved if it is **mounted vertically.**

Yacht to yacht: range 15–20 miles.

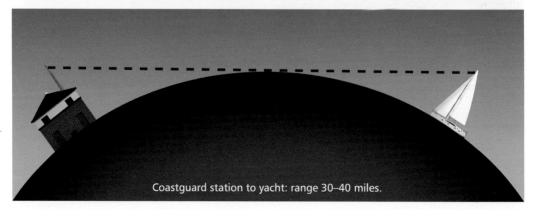

Coastguard station to yacht: range 30–40 miles.

Portable to portable: range about 5 miles

It should be possible to talk to a Coastguard station from **30 to 40 miles** offshore because of the height of their aerial.

Transmitting range is also affected by the **transmitting power of the set.** The maximum power allowed is **25 watts.** There is also a low power setting, which reduces the transmitting power to **1 watt** for routine communications. You might think that it is always a good idea to broadcast your signal as far as possible. This it not so. Remember that each channel can only be used for one transmission at a time. Powerful signals cause more inference to other radio users. If you are calling another craft nearby or a marina, use low power. Try to use low power for all routine communications. The use of low power does not change the receiving range of the set.

All distress calls should be transmitted on high power.

A portable VHF set has yet another type of aerial. This is flexible and will operate at a wider range of angles. The low aerial height and a maximum power output of 5 or 6 watts reduces the range of transmission of these sets. Between portable radios the range can be up to 5 miles, if there is no land in the way! Remember, with portable radios there is always the risk that the battery will go flat.

The information about ranges of transmissions is for average conditions and good circumstances. Ranges can be influenced by:

• Atmospheric conditions, especially high pressure, can increase the range and cause interference from distant stations.

• Land. Boats operating near land may have poor reception with signals being blocked by hills or buildings.

• Incorrect installation of the aerial, or damage to the coaxial cable connecting the aerial to the set, can give poor reception.

• The proximity of other electronic equipment can cause interference.

For these reasons it is best to have the fitting done, or at least checked, by a professional electronics engineer.

Many yachts carry an **emergency VHF aerial** in case of dismasting, which is a very good idea, but failure of the electrical supply is a more frequent problem! The emergency aerial has a plug attached to connect it to the set. For maximum range, situate the aerial as high as possible, but realistically expect a greatly reduced range.

When the mast is lost, many people are surprised to hear the radio apparently still working. This is because the coaxial cable is acting as an aerial over a short range, but transmitting without an aerial will damage the set permanently. A portable radio could be useful under these circumstances!

4 How to begin using a VHF radiotelephone

A VHF radiotelephone (be it a basic set, a transportable, or a VHF-DSC) is basically a radio with a microphone, so the controls are not complicated and often similar to an ordinary domestic radio. For example, it has an on/off switch, volume control, channel select etc.

1. Volume/On/Off
These control knobs are often combined. Switching on the set will usually bring up the display to show the settings of the other controls. It is usually possible to illuminate this at night.

2. Channel select
The method for changing channels varies with different sets and is a matter for individual choice. Key pads and scroll buttons are probably the most common. Modern sets have all the channels programmed in and the channel selected will be shown on the display. Most sets will automatically select channel 16 (and high power) when switched on. This is because **channel 16 is the international distress channel,** used for calling only when there is no alternative.

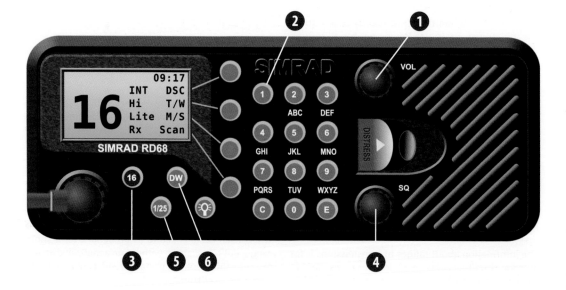

3. Channel 16 button

Channel 16 is so important that there will be a dedicated channel 16 button, often **red** in colour. This is to make it immediately obvious how to select channel 16 in an emergency.

4. Squelch

Adjust the squelch control to give the best possible reception. It acts like a filter to reduce the background noise. **Turn the squelch** until the background interference is heard, then turn it back just a little. Take care not to turn it back too much or so much signal will be filtered out that virtually none will get through!

5. High Power/Low power

On the power switch there are only two possible choices: high or low, **25 watts or 1 watt.** When switched on most sets will automatically select high power because this is the correct setting for a distress call. The power switch changes only the transmitting power and therefore range of the set. Routine calls to a marina or vessels in the vicinity should always be made on low power as this reduces the interference to others. Remember, any use of the radio occupies that channel and blocks other communications. This is why all calls should be brief, on the correct channel, **using low power whenever possible.**

6. TW (Tri-Watch) or DW (Dual Watch)

The use of this facility enables the set to **monitor** two or three channels at the same time. This will be channel 16 and one or two others. The set will give priority to channel 16. Select the second channel, e.g. the operations channel of the local port, then press the DW button. On the screen it is usually possible to see the numbers switch back and forth as the set monitors both channels. If you want to transmit turn off the DW and select the correct channel as not all sets do this automatically. Some sets can **scan** through any or even all the channels in succession. When scanning like this there is no priority channel. Read the instruction book to find out how to do this, should you want to. You may find you hear nothing clearly on any of them!

Microphone Some sets come with a fist microphone and others with a telephone handset. **A fist**

microphone will have a PTT or 'press to talk' button on the side. The microphone should be stored carefully when not in use so that the button is not pressed accidentally. If this happens the set will transmit and occupy the channel. (The channel is occupied whenever the button is pressed, even if no one is speaking. Everyone will hear sounds, and even conversation, on the boat. All this will cause serious disruption, probably on the emergency channel, and major embarrassment too.) A newer development is the command microphone which has the ability to adjust the controls of the set remotely.

Before transmitting, work out what you want to say, maybe write a few notes, and then listen for a few minutes to be sure that your call will not interfere with other traffic. Hold the microphone about 2 inches away from your mouth, press the PTT button and then start speaking. It is important to speak **very slowly and clearly**. Speak in **short simple sentences** and when finished say 'over' and release the PTT button. This is extremely important. The set can either transmit or receive, not both at the same time. If the PTT button is not released you will not hear an answer.

The reply will be heard from the loudspeaker incorporated in the set. An additional external speaker can be fitted in the cockpit or on the bridge – very useful at sea but remember to switch it off in marinas!

With a **telephone-style handset** the procedure is very similar. The PTT button is usually on the inside of the handgrip to be convenient for fingertip control. When the handset is removed from its holder it may cut off the internal speaker so the reply comes into the ear, just like on a phone. This handset needs to be carefully replaced after use or the internal speaker may not be switched on again, and so no signals heard. On the screen there is usually something to indicate when you are transmitting. A light may come on or the symbol 'RX' (receiving) may change to 'TX' (transmitting) when the PTT button is pressed.

If a **fault** develops with the PTT button the radio might be transmitting and occupying the channel. This is not common but is very disruptive. The operator can check for this by:

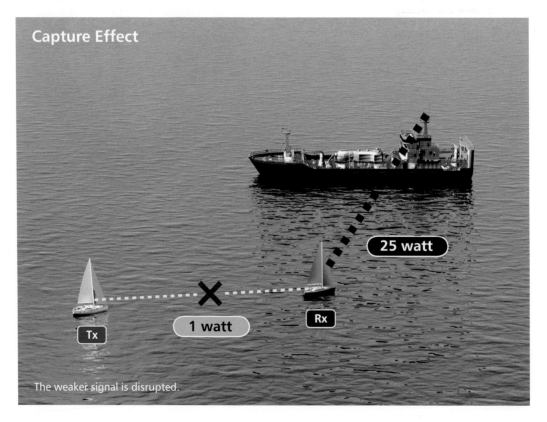

Capture Effect

25 watt

Tx

✗

1 watt

Rx

The weaker signal is disrupted.

- Looking for the 'TX' symbol or light on the screen

- Being aware that the set is not **receiving** any normal transmissions

- Testing with the squelch control. Even if it is turned fully, no squelch is heard.

- Pressing and releasing the PTT switch on the microphone **once** to hear a 'click'. To avoid interference, **do not use channel 16 for this test.**

The Capture Effect

As soon as the VHF is switched on it is in receive mode and transmissions can usually be heard, sometimes with one message being disrupted by another call. This is because the most powerful signal will override the weaker ones in what is called the **capture effect.**

If two vessels are communicating on low power when a more powerful transmission starts the boat in receive mode will only hear the more powerful signal. The message is disrupted and the situation can be very confusing, especially for the vessel transmitting. On the transmitting vessel the PTT button is pressed, therefore the radio is unable to receive, and the operator is unaware that their message has been obliterated. I have heard receiving stations asking for clarification use the proword 'over spoken', as in "Say again, you were over spoken", which exactly explains the situation.

To avoid causing this problem for others, use low power for all routine calls, keep each transmission as brief as possible and use the fewest number of interchanges to minimise your use of any channel. The capture effect is **not** a good reason to use high power, except in distress, it is one of the causes of the problem!

5 The DSC Controller

If the set has a DSC controller it may be a separate unit or more commonly incorporated into the set. Whichever is the case there will be some controls **in addition** to the ones on the basic set. **A Class D controller** is the type designed for use on yachts and motor cruisers.

The controller sends on **channel 70** a digital burst of transmission lasting about 0.5 seconds to initiate a distress alert or indicate to a selected station that you are calling them. It has been likened to a **pager**. The receiving station is informed of the incoming message by an alarm, and the details will follow. This is what the DSC part of the set does. Once an alert has been sent to a particular vessel or to all vessels in the area the voice message should be sent in the normal way. The DSC is usually associated with sending distress calls but it is used for routine calls as well.

The controller has a display showing the additional features. The **position** and **time** from the GPS may be shown if the set is interfaced with a GPS via an NMEA connection (marine industry standard for interfacing). Most GPS sets are capable of this, even the small, basic ones. The **MMSI** of the vessel may be shown too, if not write it by the set.

One of the most obvious features of the set is a **red button** labelled **distress alert** or **SOS**. This button must have a cover to prevent accidental activation. The **cover** has to be opened and then the button pressed for 5 seconds to send the distress alert. The alert is sent on channel 70, sounding an **audible alarm** on all DSC sets within range. This alert will be retransmitted **about every four minutes** until acknowledged by a class A or B controller at a Coastguard station or on a ship. (A class D controller is not capable of digitally acknowledging a distress alert.) The voice distress message on channel 16 should follow the distress alert. A skipper without a DSC set would hear the message on channel 16 but not the initial alert. When the controller is used to initiate a call the digital signal on channel 70 will always include the **MMSI**. A distress alert will also include the **time and position** (if the set is linked to GPS) and can indicate the nature of the distress. All this in 0.5 seconds! To indicate the type of distress the user can scroll through a choice of ten before sending the alert. Alternatively the set will default to 'undesignated' or '**undefined**'. If no GPS information is available **the position and time can be entered manually**. This is not quick and mistakes are easy to make. It is recommended that the position be

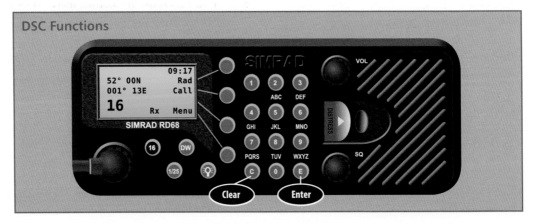

DSC Functions

updated every hour and after 4 hours the set may 'prompt' for this. If it is more than 23 hours out of date it will not be sent. The radio cannot update the position itself. The value of the set in an emergency is greatly reduced if it is not linked to GPS. **Automatic interfacing is highly recommended**. Remember, it is only with a distress alert that all this information will be transmitted by the controller.

The **MMSI** will always be transmitted by the controller:
• These are like telephone numbers for **vessels** and other **radio stations**.
• They provide unique identification.
• All MMSIs have 9 numbers.
• The first three numbers are known as the **Maritime Identification Digits** (MID). They identify the nationality of the vessel.
• All UK vessels have an MMSI starting **232, 233, 234** or **235**.
• All coast stations have an MMSI starting **00**.
• In addition to an individual MMSI a group MMSI number can be allocated by Ofcom to a yacht club or charter fleet and put into the memory so that the alert goes to all in the group. A group MMSI begins with a single 0. This number can be changed as necessary.

This fits together like this:

The yacht Sierra	MMSI 234780510
Thames Coastguard	MMSI 002320009
Vessels sailing from a UK club	MMSI 023308325

If you want to make a call using this facility then you have to know the MMSI of the other vessel. For those not technologically challenged there is a service on the internet set up by the ITU called **MARS** (maritime mobile access and retrieval system). This is really true though I will understand if you refuse to believe it! Address **www.itu.int/MARS**

The controller has a memory and built-in **directory** so the MMSIs of friends' boats and your local Coastguard station can be stored to make setting up a call quicker and simpler. Many phones, especially mobiles, are just the same.

HM Coastguard MMSI Allocations	
002320001	MRSC SHETLAND
002320004	MRCC ABERDEEN
002320005	MRSC FORTH
002320007	MRSC HUMBER
002320008	MRCC YARMOUTH
002320009	MRSC THAMES
002320010	MRCC DOVER
002320011	MRSC SOLENT
002320012	MRSC PORTLAND
002320013	MRSC BRIXHAM
002320014	MRCC FALMOUTH
002320016	MRCC SWANSEA
002320017	MRSC MILFORD HAVEN
002320018	MRSC HOLYHEAD
002320019	MRSC LIVERPOOL
002320021	MRSC BELFAST
002320022	MRCC CLYDE
002320024	MRSC STORNAWAY

The controller can send an '**Urgency Alert**' or '**Safety Alert**' to all DSC sets. Access to these facilities and others, including the directory and log of calls received, is via the individual menu system of the set. The **ENTER** or **SEND** button will send the alert on channel 70, except the **distress alert**, which is initiated by the **red button**. Some DSC sets have additional features such as polling and position request making it possible to check if another vessel is within radio range or their GPS position, at the press of a button.

Becoming familiar with these functions will take time. Practising with a VHF-DSC switched on could lead to prosecution if an alert is sent accidentally. **NEVER TEST THE SET BY INITIATING A DISTRESS ALERT.**

On RYA courses a computer simulator (produced by Lightmaster Software Tel: 020 8405 8200) is often available. This programme can be also bought and used at home.

6 Which channel do I use?

The channel numbers that we see on the screen are pre-set with the frequency the radio will use when transmitting and receiving. By international agreement all marine transmissions take place between 156.00 MHz and 174.00 MHz. There are 59 channels within this band and they are allocated for different uses. **You need to know some of these channel allocations**; for others it is sufficient to know where to find the information. In all cases the important thing is not to use a channel unless you are sure that it is available. Just because there seems to be no one talking does not mean that the channel can be used. Using the incorrect channel can cause serious interference.

Channels are allocated for different uses. Sometimes one channel is used as the **calling channel** and then another is used to pass the message. The latter is known as the **working channel**. Whenever possible call directly on the working channel, or use DSC.

Which channel to use?

STATION CALLED	CALL ON CHANNEL	WORKING CHANNEL
Coastguard	DSC or 16	67
Solent Coastguard	67	67
Ships/boats	DSC or 16 unless pre-arranged	6, 8, 72, 77
Port	Look up	
Marina	80	80

Channels are allocated to specific uses under international agreement.

Distress and safety: channel 16.
This is used for distress communications. It is the channel that should be **monitored** while at sea. By listening to this channel the skipper will also hear Maritime Safety Information broadcasts from the Coastguards. These include regular weather forecasts as well as gale and navigation warnings. Following the announcement the information will be read on another channel. **Avoid using channel 16 as a calling channel whenever possible,** by using DSC, prearranging a channel, calling on the working channel or using a mobile phone. No transmission on channel 16 should be longer than **one minute unless it concerns distress**. Never pass a routine message, however brief, on channel 16. If a call goes unanswered wait **two minutes** before calling again, except in a distress situation.

Digital Selective Calling: channel 70
This is the channel that the DSC controller uses. **It is illegal to use channel 70 for voice transmissions**.

Intership: channel 6, 8, 72 and 77
Interships channels are used for communications between yachts, boats and ships. There are many channels listed as intership, but 6,8,72 and 77 are recommended for small craft. Channel 6 is designated as the primary intership channel and must be available on all VHF sets. It may occasionally be necessary to call a vessel on channel 16 but the message should be passed on an intership channel. It is a very good idea to call on the intership channel if possible. If several vessels from a club are on passage they could nominate an intership channel to use, perhaps channel 77, and then listen on dual watch to channel

16 and 77. If a call is made it should be made direct on channel 77 thus keeping channel 16 clear for any distress communications. This arrangement needs to be definite and adhered to by all the skippers. If a VHF set is monitoring two channels on dual watch, unless someone happens to be looking at the screen at the time, it is not possible to tell on which channel the call was received. DSC always avoids the use of channel 16 for routine calls.

Bridge to bridge communications: channel 13
This channel is for vessels in confined waters or shipping lanes to communicate on matters of navigational safety. Monitoring this channel is a good idea. If it is necessary to contact a ship, try channel 13.

Public correspondence:
numerous channels
These channels are for the use of Coast Radio Stations, which link the VHF with the telephone system and allow calls to subscribers ashore. This was an expensive service, which has now ceased in the UK. Information is available in **Almanacs** or **The Admiralty List of Radio Signals** on any stations still available. The Admiralty now publish a volume of the ALRS especially for small craft, NP289, which is available from chart agents.

As CRS channels become redundant they are being reallocated for other purposes.

Port operations and ship movements:
numerous channels
Channels 12 and 14 are commonly used for port operations. The channel used by each port is listed in the almanac and ALRS and sometimes shown on the chart. These channels are used for co-ordinating the ships in the Port Authority areas.

Many ports, especially those busy with commercial shipping, require all vessels to call on the working channel and ask permission to enter or leave the harbour. Details of this will be in the almanac too. Even if this is not mandatory, monitoring the channel could be very interesting and useful. Ship movement channels tend to be used for communications between tugs, shipping offices and ships.

UK ONLY

Small craft safety communication with the Coastguard: channel 67
This is the channel most commonly used by the Coastguard to talk to small craft, following a DSC call or call on channel 16. In the area covered by **Solent Coastguard** channel 67 should be used as the calling channel, if DSC is not available.

Marinas: channel 80
Use Channel 80.
All skippers should be familiar with this as it is the same channel for all marinas, unlike ports. There tends to be plenty of interference and confusion on channel 80 especially at teatime on Saturdays! To help as much as possible check before calling that there is not another call in progress, use low power and listen to be sure that the reply you hear is intended for your vessel. You can help the radio operator at the marina by saying the vessel's name clearly. **Never call a marina on channel 16**. They do not have channel 16!

Yacht clubs: channels M and M2
Channel M2 should be the first choice channel for yacht club use when organising racing safety boats or water taxis. Channel M can also be used.

Private channels
There are some frequencies available for private channels. A few are used exclusively by the rescue services but others can be allocated to a private organisation like a sailing school, charter fleet or tug company on the payment of a special licence fee.

Simplex / duplex

Simplex / duplex and why sometimes you cannot hear both sides of a conversation.

At the back of the book, on pages 68-69, is a list of all the radio channels and the frequency, or in some cases the two frequencies, allocated to them under international agreement. As normal users of the radio we do not need to know these frequencies, because for each channel these have been pre-tuned into the set. Channels with one allocated frequency are

Simplex channel - one frequency.

Duplex channel - duplex working.

simplex channels and those with two are **duplex** channels.

A **simplex** channel with only one frequency allocated to it requires the two radio operators to take turns in speaking, making the conversation like a game of "ping pong". Each operate has to wait for the other to stop transmitting, by releasing the PTT button, before they can reply. The radio switches from transmit to receive as it is released. No reply, or other transmissions, can be heard until this happens. The signal to the listening operator that it is their turn to speak is the proword "over", from the original "over to you, over".

On a **Duplex** channel there are two frequencies, one to transmit and the other to receive. The radio operator speaks on one frequency and listens on the other, making the conversation like a telephone. To achieve this normal two way conversation both radio stations require two aerials. In some countries Coast Radio Stations use duplex channels to link VHF operators with telephone users ashore as part of their telecommunications system. This is not available in the UK.

For most small boats full two way duplex conversations, are not possible because of the single VHF aerial, but duplex channels can still be used with limitations. Using a duplex channel with a single aerial is known as **semi-duplex**.

In **semi-duplex** working the boat is restricted to transmitting or receiving through the single aerial, but the radio automatically uses one frequency to transmit and the second frequency to receive. For the radio operator there is no real difference between this

and using a simplex channel, but one slightly strange consequence of semi-duplex working is that other boats in the area **cannot hear both sides of the conversation**. I have heard people comment "Well if I cannot hear it why does it matter?", and I have some sympathy for this point of view, but there is an impact on us all in the case of **channel 80**, which is a duplex channel. Channel 80 tends to get busy at particular times of the day, especially if there are several marinas in the same area and because it is a duplex channel each caller **cannot hear all the other boats calling**. They are unaware of how busy the channel is and how repeated calling may be causing mutual interference.

Another interesting complication that can occasionally occur is due to differences in these arrangements between countries, the most significant for us being that channel 80 is duplex internationally, but simplex in the USA. You don't have to sail there or charter in the Caribbean to be affected by this because modern sets are manufactured to be sold in many countries

Duplex channel - semi-duplex working. All boats hear the reply.

Only marina hears the calls.

and are capable of being used on either international or US channels. A set bought over the internet may be programmed inappropriately or an owner can change it accidentally. This may go unnoticed until channel 80 is used to call a marina. The marina will hear the call on the receive frequency of channel 80 and reply on the transmit frequency, but the boat will not hear this reply. Their radio will not have switched from the transmit to the receive frequency as it should when the PTT button is released if it is using US channels. This is not a common problem but if calls to a marina go unanswered, and you are definitely within range, check on the screen. It should say **Int** and not **US**. If in doubt read the instructions, and remember sets imported from the US may not be legal in the UK or EU.

Prohibited Calls

Some types of calls are not allowed:

- An unqualified person, unless under the control of a VHF or SRC certificate holder, may not use the radio.

- Calls unidentified by a vessel's name, call sign or MMSI, may not be made. The name of the operator cannot be used instead.

- Calls not authorised by the skipper are prohibited.

- No false or deceptive distress signals should be sent.

- Turning off the set after a distress message has been sent but before the situation has been resolved is not allowed.

- The broadcasting of music or other messages is prohibited. (Broadcast in this sense means to transmit speech or music on a radio channel without expecting a reply as the pirate pop radio stations used to do.)

- Calls should not be made to stations ashore except ports, coastguards and marinas and should be on the channels covered by the licence.

- Profane, indecent or obscene language is not permitted

- Making unnecessary transmissions or transmitting superfluous signals is not allowed.

The phonetic alphabet

On the radio everything must be said very slowly, clearly and as briefly as possible. The phonetic alphabet helps. It is an internationally recognised method of spelling out difficult words, especially names. It is also used for the call-sign. All radio operators need to know the phonetic alphabet. Before spelling a word, say the word and then "I spell" to warn the other person to be ready.

A	alpha	B	bravo	C	charlie
D	delta	E	echo	F	foxtrot
G	golf	H	hotel	I	india
J	Juliet	K	kilo	L	lima
M	mike	N	november	O	oscar
P	papa	Q	quebec	R	romeo
S	sierra	T	tango	U	uniform
V	victor	W	whiskey	X	xray
Y	yankee	Z	zulu		

7 What do I say? ...Prowords

Prowords are procedural words that have a commonly understood meaning and are intended to make communications brief and easily understood. **Avoid jargon** that is heard in old TV programmes and just sounds silly. Avoid questions like "Are you receiving me?" because if the call is not heard then no one will reply. You will know the answer! "How do you read?" is a question from the time that old military radios had dials to show the level of reception.

These are a few of the most useful prowords used in normal calls.

Proword	Use	Example
Say again	To ask for a repetition	"Say again, which channel?"
I Say again	To give a repetition	"I say again channel seven seven"
All after	To obtain clarification	"Say again all after ..."
All before	of part of a message	"Say again all before ..."
Correction	Said before correcting part of a message	"Go to berth 273. Correction berth 275.1 say again, 275."
Spell	Said before use of the phonetic alphabet to spell a word, usually a name	"Vessel Indigo. I spell – India-November-Delta-India-Golf-Oscar."
This is	To identify calling station	"This is yacht Indigo."
Over	Invitation to reply as the PTT switch is released	"This is yacht Indigo, over."
Out	The end of the conversation. No reply expected	"Thank you, out."
Station calling	Used if the name of the vessel calling is not understood	"Station calling Sierra. This is Sierra. Say again your name, over."
Stand by	Used if the message cannot be exchanged immediately	"Solo. This is Thames CG. Channel 67 and stand by."

From this table, and using common sense, you can see how short phrases can be put together in recognised patterns to make the meaning clear and to keep all messages as brief as possible. Several short statements are more easily understood than a long rambling sentence. Avoid unnecessary repetitions and speak slowly. Remember that someone may be trying to write down what you are saying.

Prowords used in Distress situations

MAYDAY Under international law the word Mayday may only be used if **"there is grave and imminent danger to vessel, vehicle, aircraft or person which requires immediate assistance"**. It is vital to remember this. The situation must be life threatening. Every captain and every skipper is obliged to provide assistance in a Mayday situation if they can. They should not put their vessel or crew in danger but they must help if possible despite inconvenience to crew, passengers or shipping company. Sending a Mayday or being involved in assisting in a Mayday situation is a serious responsibility and can only be undertaken on the skipper's authority. The origin of Mayday is the French "m' aidez", meaning "help me".

Seelonce Mayday. While a Mayday is in progress radio silence is automatically imposed, on the distress channel, on all vessels not involved. The Coastguard dealing with the incident or the casualty vessel may say "Seelonce Mayday" if it is necessary to remind shipping of this.

Seelonce Distress. Any station nearby who is convinced that it is necessary to impose radio silence can use the expression "seelonce distress".

Prudonce. Complete radio silence may not be required once the rescue services are involved and are dealing with the incident. "Prudonce" is used to relax but not lift the radio silence. The VHF can be used prudently. Any channel being used for communications concerned with the distress situation should be used with extreme brevity.

Seelonce Feenee. Radio silence is lifted.

Mayday Relay. A message sent by a vessel, not itself in distress, passing on the distress information. This would be done if the distress vessel could not transmit the Mayday or has indicated its distress with flares or some other internationally recognised distress signal. A relay might be used if a Mayday was heard which did not seem to be acknowledged by the Coastguard or any other vessel which might be able to assist. **This is unlikely in an AI area under normal circumstances, but** the casualty could have a problem which drastically reduces the range of their transmissions or be in a location which is a radio 'blind spot'. It is important that the separate identity of the casualty vessel and the vessel sending the relay are made clear.

Received Mayday. Used, usually by the Coastguard, to **acknowledge** receipt of a distress message. This implies that assistance of some kind will be provided. This could be direct assistance by a vessel nearby or indirectly provided as in the case of the Coastguard sending a lifeboat or helicopter.

Pan Pan. Pan Pan is used to preceed an urgent message when the skipper may need some assistance, but where the situation is not life threatening. This might be an engine failure, where the boat is not in immediate danger, but does require an urgent tow.

Pan Pan can also be used if the urgent message concerns a request for medical advice or assistance. The origin of pan pan is from the French "en panne" meaning broken down.

Securite. A message containing safety information can be preceded by the word securite. This is most commonly used by the Coastguard before broadcasting a new gale or navigation warning.

8 Ship-to-Ship routine communications

Transmitting unnecessary signals is not permitted on VHF radios. This means that the radio should not be seen as a method of 'chatting' between boats. Mobile phones work well on boats near the coast and could be used for arrangements to meet at the pub etc. If the radio is used for passing information or asking questions between craft then the calls should be brief and to the point.

Remember that if just two boats are holding a conversation then the VHF channel is occupied and no other communication is possible on that channel. Remember too that it is a huge party line...anybody might be listening. Using low power when the other boat is nearby will at least limit the number of people who have to wait for you to finish before they can make their call.

To make a **routine call** to another craft using VHF-DSC you need to know its **MMSI**. The controller has the facility for a **directory of MMSIs**. These are programmed in by the owner to include those numbers regularly used. An audible alert will sound on the receiving set, which will then change to the intership channel nominated by the calling DSC when the call is answered. The controller maintains a **record of calls received** in the same way as a mobile phone. It is the operator's responsibility to choose a correct intership channel. Those recommended are **6, 8, 72** and **77**. DSC is great for these calls as it is selective, avoids channel 16 and the receiving set goes to the chosen channel automatically. If not using DSC the initial call may have to be made on channel 16, and **must include a suggested intership channel**. If arranged in advance it is both possible and sensible to call direct on the intership channel. **Avoid using Channel 16 as the calling channel whenever possible**.

Group Calls

Another special feature of the DSC set is the ability to make a **group call**. A fleet of boats from a club or a sailing school can request Ofcom to allocate a **group MMSI**. This number can be programmed into the set in addition to the individual MMSI. The number is instantly recognisable as a group MMSI because it begins with a **single zero**, followed by the country code and 5 digits. When this MMSI is used in a routine DSC call the alarm for an incoming call sounds on all the boats in the group and all the sets automatically retune to the intership channel nominated by the calling vessel, when the call is accepted.

With a non-DSC set the group could make an arrangement before leaving the marina which intership channel to use. This channel would then be monitored, together with channel 16, using the **dual watch facility.** For one boat to communicate with the group, it should de-select dual watch, and call on the intership channel. This is an extremely efficient method of working.

DSC Routine Call

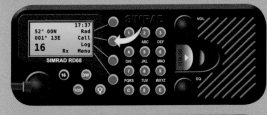

1 Select **call**.

2 Enter **MMSI**.

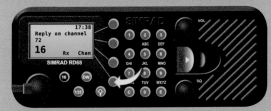

3 Select **Intership Channel**
   – 6,8, 72 or 77.

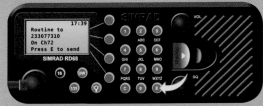

4 Press **Send**, then E to confirm.

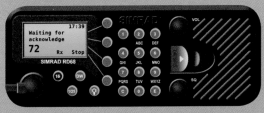

5 Wait for acknowledge then transmit
   using low power.
   "Swift. This is Indigo.
   Our ETA at the marina is now 17.00.
   Over."

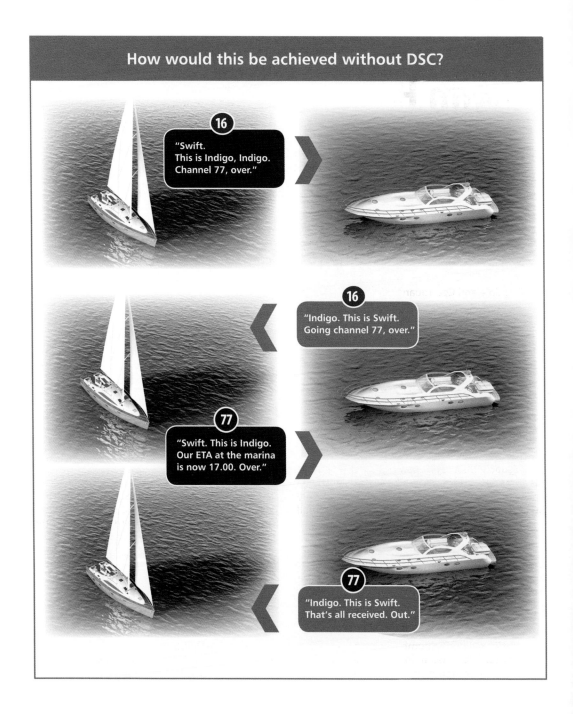

9 Routine communications with HM Coastguard

The Maritime and Coastguard Agency (MCA) is the full and correct name for the government body responsible for co-ordinating search and rescue in UK waters and marine safety. It was formed in 1998 by the merging of HM Coastguard and the Marine Safety Agency (MSA). The MSA was largely seen as the body responsible for the safety and regulation of shipping, but they also regulated standards of liferafts, lifejackets and other safety equipment that we all use, and which are mandatory on sailing school and charter boats.

Those involved in leisure boating know the Coastguard best by listening to the regular Maritime Safety Information broadcasts and even get to recognise individual voices. **It should only be under exceptional circumstances that a small boat would need to call up for information that is included in the MSI broadcasts.**

Owners can send information about their craft to the local Coastguard station by joining the **Voluntary Safety Identification Scheme.** The card, often called the **CG66,** should be filled in and sent to the nearest Coastguard station. This gives details of the vessel, its safety equipment and contact telephone numbers. In an emergency these details can assist in a rescue. The card needs to be updated every two years or sooner,

if necessary. These cards are available from your local Coastguard station, by telephoning the MCA on 023 8032 9100 or the form can be completed online at **www.mcga.gov.uk.**

Vessels can **log details of a passage** with the local Coastguard station. It is important for the officer to have the correct spelling of the name of your craft in case a computer search for mention of your boat is made in an emergency, or if the vessel is reported overdue. All radio calls they receive are logged on computer so they may have to ask you to spell the name of the boat if they are in any doubt. Naturally you should use the phonetic alphabet!

Call using DSC and the MMSI of the individual Coastguard station. It's a good idea to enter the MMSI of the nearest Coastguard station into the directory in the set. If necessary, call on channel 16 and request channel 67, unless you are in the area covered by Solent Coastguard when you should call direct on channel 67.

Remember to call and report your safe arrival. Lifeboats and helicopters will not be launched if you forget, but it is correct procedure. If your destination is out of range of the original Coastguard station then call the nearest Coastguard station and they will log your arrival.

Routine call to the Coastguard using DSC

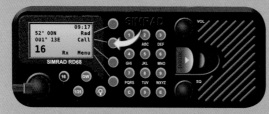

① Select **call**.

② Scroll through directory to **Thames Coastguard**.

③ Press **send**.

④ Press **E** to confirm.

⑤ The set will switch to channel 67 when the call is acknowledged by the Coastguard. Transmit on low power. "Thames Coastguard. This is Sierra. MMSI 234780510. Over."

Routine call to the Coastguard using Channel 16

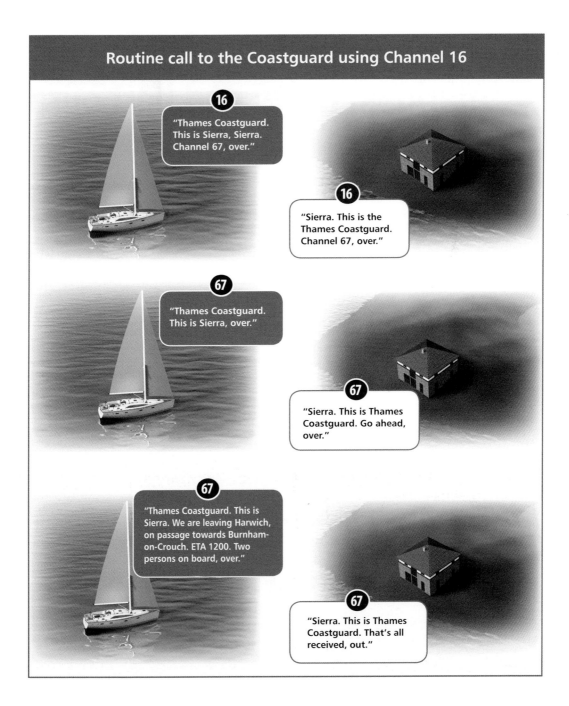

10 Routine communications with marinas, ports and harbours

The only channel for calling a marina is **channel 80.** Marinas do not have VHF-DSC. In a popular cruising area therefore all the marinas are using the same working channel and tend to have the same busy time for calls. You can help by using **low power,** listening before you start your call to ensure that you do not interrupt others, and by being as brief as possible. With this giant party line, calls can get muddled and instructions confused.

Calls are made by craft requesting a berth, either overnight or for a short stay at lunchtime; skippers wanting diesel; general assistance or information about lock openings. If requesting a berth be ready with the length and beam of the boat and say immediately if the berth needs to be for more than one night. If requesting diesel then follow the local routine: this may mean making the call once the boat is alongside the diesel berth. This sort of radio work is excellent practice. Keep the messages short and clear, saying the name of the boat especially carefully.

Visiting boats often use the spaces that resident berth holders who are out for the day or on a cruise have left empty. For this reason it is in the interests of all marina berth holders to say when they will be away and when they expect to return. The marina is always pleased to have this information, especially when demand by visitors is high.

A brief radio call on channel 80 to tell the staff that the boat will be away or from a visitor to say that the berth is now vacant again is an excellent way of checking the radio is operating correctly!

Routine call to a marina using channel 80

80
"Suffolk Yacht Harbour. This is Solo. We are alongside the fuel berth. May we have some petrol please? Over."

80
"Solo. This is Suffolk Yacht Harbour. Good morning. Someone will be down as soon as possible."

A **radio check** with the Coastguard on channel 16 is still popular with some 'dinosaurs' but it is **not recommended** and not really necessary. The radio is very likely to be working correctly, especially if it was last weekend, and radio checks just clutter up the airwaves. Only call the Coastguard for a radio check **if it is absolutely vital and there is no alternative.** A radio check in that case, if it involves channel 16, should last no longer than **10 seconds.**

Ports and harbours

Port operations for large commercial ports are handled from an operations room, like a shipping version of air traffic control. They are busy and the radio traffic is vital for the safety of crew and passengers.

The radio communication for each port is handled on one of the **port operations channels.** Ports use different channels from the other harbours nearby to avoid interference. They monitor channel 16 but radio traffic is handled on the **working channel.** As a call-sign some harbours now use **VTS** as in 'Harwich VTS'. This stands for vessel traffic service. In naval ports it might be necessary to address the call to the 'Queen's Harbourmaster' (QHM).

The working channel will be listed in the almanac, and sometimes on the chart. There too will be information about a special navigation channel to follow, local sound signals or regulations. These local instructions may include the requirement to use the engine or **to call on the radio for permission before entering or leaving the harbour.** This is most common where the navigation channel available for commercial traffic is narrow and shared with small craft. At these ports you can expect strict traffic control for pleasure boats as well as commercial shipping. In these congested ports the instructions from the harbour radio must be obeyed even if the reason is not immediately obvious. Other ports do not want the small boat skipper to call except in exceptional circumstances. This is often where there is more space and perhaps separate navigation channels for commercial and leisure craft.

Ports do have VHF-DSC equipment but this does not change the advice about monitoring the working channel for information about shipping movements and calling on that channel should it be necessary.

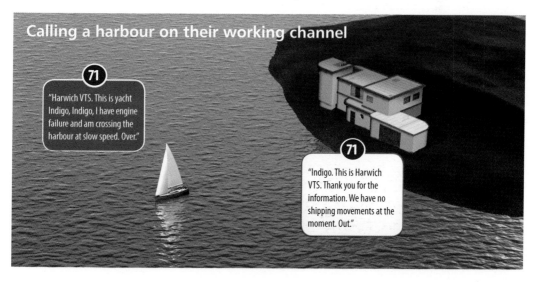

Calling a harbour on their working channel

71
"Harwich VTS. This is yacht Indigo, Indigo, I have engine failure and am crossing the harbour at slow speed. Over."

71
"Indigo. This is Harwich VTS. Thank you for the information. We have no shipping movements at the moment. Out."

11 On passage with yacht Sierra radio in everyday use

Sierra makes a passage from Suffolk Yacht Harbour to Burnham-on-Crouch.

On this relaxed and peaceful trip Sierra uses the VHF-DSC radio for routine communication, to receive information broadcast by the Coastguard and to monitor other channels for safety reasons. Use low power for all routine calls.

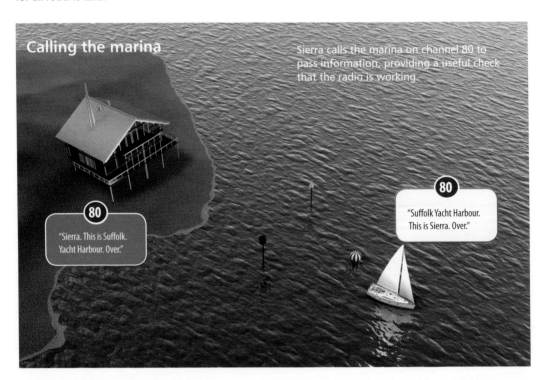

Calling the marina

Sierra calls the marina on channel 80 to pass information, providing a useful check that the radio is working.

80
"Sierra. This is Suffolk. Yacht Harbour. Over."

80
"Suffolk Yacht Harbour. This is Sierra. Over."

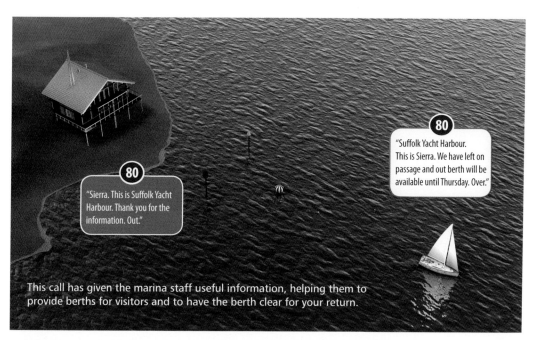

"Suffolk Yacht Harbour. This is Sierra. We have left on passage and out berth will be available until Thursday. Over."

"Sierra. This is Suffolk Yacht Harbour. Thank you for the information. Out."

This call has given the marina staff useful information, helping them to provide berths for visitors and to have the berth clear for your return.

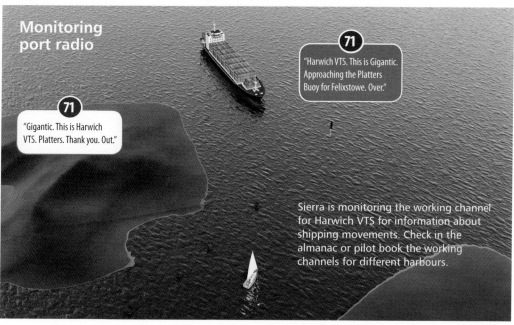

Monitoring port radio

"Harwich VTS. This is Gigantic. Approaching the Platters Buoy for Felixstowe. Over."

"Gigantic. This is Harwich VTS. Platters. Thank you. Out."

Sierra is monitoring the working channel for Harwich VTS for information about shipping movements. Check in the almanac or pilot book the working channels for different harbours.

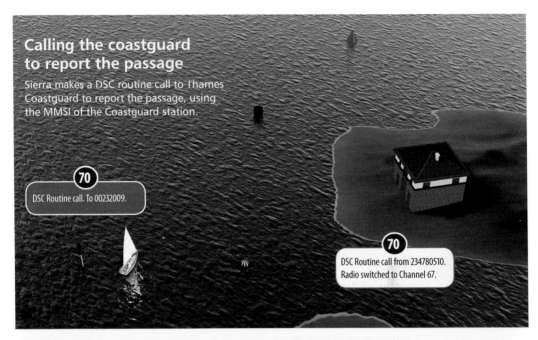

Calling the coastguard to report the passage

Sierra makes a DSC routine call to Thames Coastguard to report the passage, using the MMSI of the Coastguard station.

70
DSC Routine call. To 00232009.

70
DSC Routine call from 234780510.
Radio switched to Channel 67.

67
"Thames Coastguard. This is Sierra MMSI 234780510. We have left Harwich on passage to Burnham. 2 persons on board ETA 12.00. Over."

67
"Sierra. This is Thames Coastguard. That's all received. Out."

Monitoring a yacht club channel

There is a yacht race about to start. Sierra can monitor channel M (37) and keep clear.

M "All competitors. This is the committee boat. Stand by for the 5 minute gun. Out."

Calling – receiving no reply

Sierra calls the RIB Solo on channel 16 because his portable radio does not have DSC. Solo is out of range and there is no reply. Wait 2 minutes. Call again. If there still is no reply STOP CALLING.

16 "Solo. This is Sierra. Sierra channel 72, over."

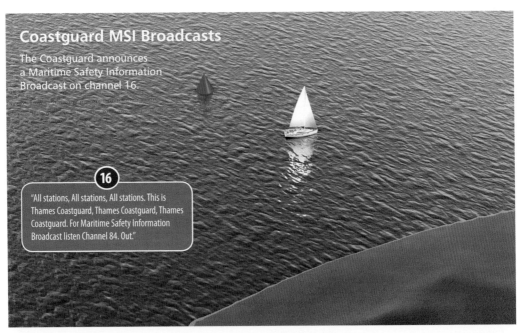

Coastguard MSI Broadcasts

The Coastguard announces
a Maritime Safety Information
Broadcast on channel 16.

16

"All stations, All stations, All stations. This is
Thames Coastguard, Thames Coastguard, Thames
Coastguard. For Maritime Safety Information
Broadcast listen Channel 84. Out."

Sierra switches to channel 84 to listen to the
weather forecast and navigation warnings.

84

"All stations, All stations, All stations. This is
Thames Coastguard, Thames Coastguard, Thames
Coastguard. Here is the weather forecast followed
by the navigation warnings."

Receiving a call by DSC

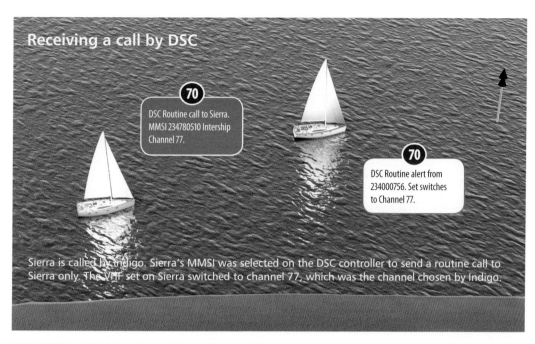

Sierra is called by Indigo. Sierra's MMSI was selected on the DSC controller to send a routine call to Sierra only. The VHF set on Sierra switched to channel 77, which was the channel chosen by Indigo.

The reply is brief and to the point.

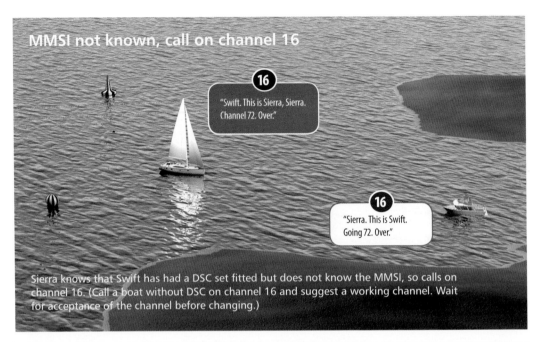

Sierra knows that Swift has had a DSC set fitted but does not know the MMSI, so calls on channel 16. (Call a boat without DSC on channel 16 and suggest a working channel. Wait for acceptance of the channel before changing.)

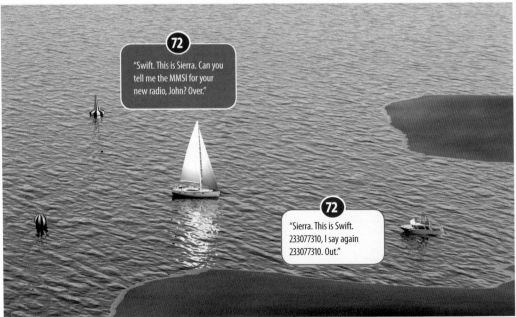

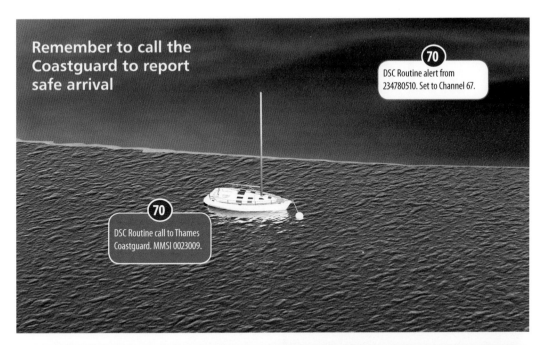

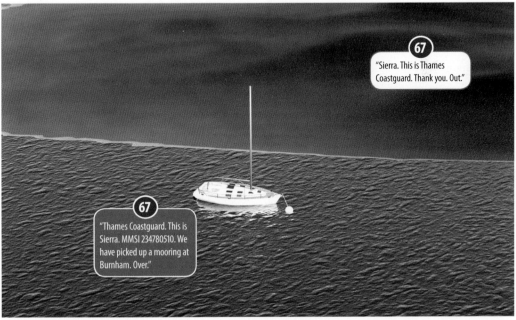

12 Distress procedures and Coastguard acknowledgement

Remember the rules concerning the sending of a MAYDAY. A Mayday may only be transmitted if there is "Grave and imminent danger", in other words risk of loss of life. It can only be sent with the skipper's permission.

The Mayday transmission is in three parts. The distress alert, using DSC channel 70, followed by the Mayday call and Mayday message by voice on channel 16. On a radio without DSC transmit on channel 16 only, omitting the MMSI:

Mayday call	Mayday Mayday Mayday This is yacht Indigo, Indigo, Indigo

Mayday message	Mayday yacht Indigo
	MMSI 234000756
	In position:
	      51°46'.50N
	      001°16'.50E
	On fire
	Require immediate assistance
	3 persons on board
	No liferaft
	Over

In distress calls the word Mayday and the name of the vessel are said three times. This is not necessary in routine calls and should be avoided.

To make the message as clear as possible use short phrases. Avoid sentences altogether and give the information in the correct order. This order is logical, giving the most important information first, and will be as anticipated by the recipient.

Say the Latitude and Longitude position very carefully and clearly. 51° 48'.5ON 001°16'.5OE

'Five one degrees four eight decimal five zero minutes North.

Zero zero one degree one six decimal five zero minutes East'.

If giving your position as a range and bearing, do it in the internationally recognised way. That is with the bearing in degrees True away from the named object, followed by the distance. It is vital to give positions correctly, and speak slowly and carefully. None of the rescue services can help unless they know where you are.

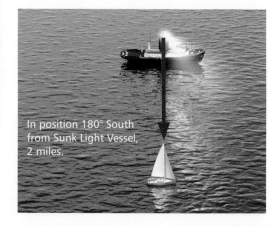

In position 180° South from Sunk Light Vessel, 2 miles.

It is important to **include the MMSI** if the Mayday voice message was preceded by a DSC distress alert so the station that receives the DSC alert knows that they are listening to a transmission from the same vessel. A vessel listening on channel 16 only will also know that a DSC alert was sent.

A Mayday procedure card is printed on page (to be confirmed) for you to photocopy, complete and put up by your radio.

The Coastguard Acknowledgement

In coastal water, an A1 area in GMDSS terms, a reply to the Mayday message will probably come from the Coastguard. This is the **acknowledgement**. Again this follows a pattern.

"Mayday Yacht Indigo, Indigo. This is Thames Coastguard, Thames Coastguard. Received your Mayday The lifeboat is launching to your assiatance. Over."

Acknowledgement

If the Mayday has been preceded by a DSC alert then all vessels within range fitted with DSC equipment and the Coastguard will receive an alarm and display of information on their controllers. Their sets will automatically switch to channel 16. All vessels in the area, not just those fitted with DSC, should hear the voice distress message.

All communications during this distress situation would normally be on channel 16 and be proceeded by **Mayday Indigo**.

"Mayday Indigo, Thames Coastguard This is the Walton Lifeboat, over."

VHF-DSC sets used by Coastguard stations and ships can acknowledge a distress alert automatically by a pressing of a button. These are class A or B controllers. This **digital acknowledgement** has the effect of cancelling the repeats of the digital alert sent on channel 70. If no acknowledgement is received the alert will be rebroadcast about every 4 minutes, even if the crew has abandoned ship into the liferaft.

If you send a DSC distress alert in error

• Switch off the set to prevent the automatic repetitions.

• Switch on again and transmit on channel 16 "All stations, all stations, all stations This is Indigo, Indigo, Indigo MMSI 234000756 Cancel my distress alert. DSC distress alert transmitted in error. Out."

DSC distress alert

1. Open the red cover.

2. Press red button.

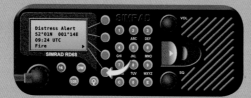

3. Select cause of distress, if time.

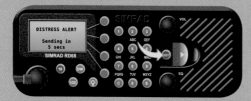

4. Press **and hold the red button through the countdown.**

5. Wait no more than 15 seconds for the acknowledgement. Send voice Mayday on channel 16 using high power.

For speed :
1. Open cover.
2. Press red button. Release.
3. Press and hold red button for 5 seconds.

13 What do I do if I hear a Mayday?

A very good question!

According to the radio regulations "it is the duty of all skippers to acknowledge" a Mayday and under international law all skippers must give assistance in a distress situation if they can do so without risk to their crew. Realistically though, in most situations a yacht or motor cruiser can do very little. Nothing must be done to confuse the situation. **The best way to help is likely to be by maintaining radio silence.** Misleading, inaccurate or confusing information can cost lives.

If close to the distress vessel and able to assist, the skipper needs to acknowledge the Mayday **only if no other station has already done so.**

If the Coastguard has already acknowledged the Mayday then they have taken control so it is important to **inform them of any action that you take.**

If the casualty is unable to send a Mayday or if the Mayday receives no acknowedgement then a **Mayday Relay** may be necessary. Remember the Mayday is extremely unlikely to go unheard in an AI area. It is possible that your vessel is out of range of the other transmissions. If a relay is required a ship will have higher aerials and other longer range communications equipment and so can relay more effectively.

It is important that all the facts are considered before deciding whether to relay, acknowledge or just listen and keep silent:

• Did the distress vessel send a DSC distress alert before the Mayday message?

If you have DSC you will know this absolutely because the alert will be visible on the screen. If not, decide on the balance of probabilities. If an MMSI was included in the voice message then a DSC was used. It is incredibly unlikely that the digital alert will not have been received by a Coastguard station. Therefore it is **unlikely that a relay need be considered** unless the distress alert is retransmitted or both the casualty vessel and your craft are outside the AI area. Any unnecessary use of the DSC alerting system, after the distress alert has already been received by the Coastguard, will cause disruption. A voice call on channel 16 would be less disruptive, if you have reasons to think that it is necessary.

• Is it possible that the casualty vessel cannot send a VHF distress message? If a distress signal, such as flares, has been sighted on a vessel or if a crew are seen abandoning their vessel then a relay could be sent.

• Might the vessel in distress have a fault on their radio or aerial, restricting the range of transmissions?

• Might they be in a location where their signal is obscured by land?

If there is cause for concern, but it is likely that a DSC alert has been sent, then call the Coastguard on channel 16.

I have just heard a Mayday... What do I do?

Position	Did you hear or see the acknowledgement?	Can you assist the casualty practically?	Action
Al Area	**Yes** from Coastguard	**No**	Maintain radio silence. Listen and write down message. Plot the casualty's position.
Al Area	**Yes** from Coastguard	**Yes**	Call Coastguard. "Mayday yacht Indigo Thames Coastguard, this is yacht Sierra. Can assist. ETA 5 minutes. Over."
Al Area	**No** (Unlikely, especially if DSC used)	**No**	If you are sure it is necessary: (DSC Urgency Alert CH70) CH16 "Mayday relay, Mayday relay, Mayday relay. This is yacht Sierra, Sierra, Sierra (MMSI) Mayday received from yacht Indigo Mayday Indigo (Repeat their message)"
Al Area	**No** (Unlikely, especially if DSC used)	**Yes**. Then acknowledge Mayday	CH16 "Mayday yacht Indigo, Indigo, Indigo. This is yacht Sierra, Sierra, Sierra. Received your Mayday Proceeding to your assistance. ETA 10 minutes." Then send a relay...

14 The Mayday Relay... 'Urgency Alert'

The purpose of the relay is **to pass on distress information** without causing confusion or interference. The first decision is whether a relay is necessary. Only relay if you hear a Mayday and it is not acknowledged by the Coastguard or another vessel, or if you see a distress signal, such as flare, or a vessel in distress that is unable to send a Mayday.

If a relay is transmitted:

• It must be clear from the transmission which vessel is in distress and the name of the vessel sending the relay.

• The facts regarding the distress situation must not get changed or muddled.

On the VHF-DSC controller the red button will send a distress alert with your MMSI and positioning. **Do not use it for a Mayday Relay**. The **call menu** makes it possible to use the DSC to send an **'Urgency Alert'** to all stations.

An **'Urgency Alert'** to all stations should be followed by a Mayday Relay by voice on channel 16. If no DSC is available transmit the relay immediately on channel 16. Remember that in an A1 area it is unlikely that the original Mayday was not received, especially if it was

DSC distress alert followed by a relay

Swift is sending a DSC distress alert followed by a Mayday call. The Coastguard cannot receive the signal because of the cliff. Sierra receives the Mayday and will do a Mayday relay.

16

"Mayday, Mayday, Mayday. This is motor cruiser Swift, Swift, Swift. MMSI 233077310. Position Black Rocks, Deadly Point. Aground and breaking up. Require immediate assistance. 2 persons on board."

preceded by a DSC distress alert. In that situation the MMSI would have been heard as part of the original distress message.

The distress alert from Swift on page TBC was sent from a location where the radio signals might not reach the Coastguard station situated on the headland above. The DSC distress alert will sound an alarm on all VHF-DSC radios within range and the message may be relayed by a ship to the coastguard. **Listen for this to avoid interference.**

If you are **absolutely convinced** that there has been no acknowledgement of the Mayday or relay sent then send a DSC 'Urgency Alert' followed by a Mayday Relay on channel 16. **If you are not sure then call the Coastguard on channel 16.**

Re-broadcast the mayday message without any changes

16

"Mayday relay, Mayday relay, Mayday relay. This is Yacht Sierra, Sierra, Sierra. MMSI 234780510. Mayday received from motor cruiser Swift. Mayday Swift MMSI 233077310. Position Black Rocks, Deadly Point. Aground and breaking up. Require immediate assistance. Two persons on board."

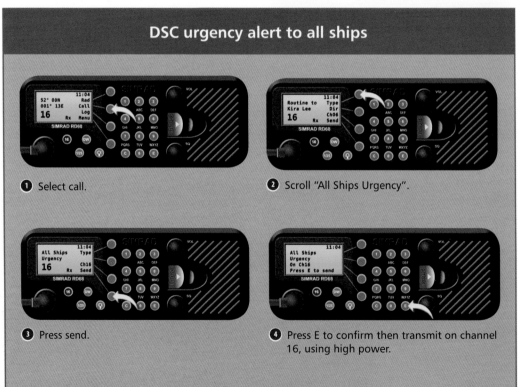

DSC urgency alert to all ships

1 Select call.

2 Scroll "All Ships Urgency".

3 Press send.

4 Press E to confirm then transmit on channel 16, using high power.

15 Pan Pan... 'Urgency Alert'

Pan Pan is the proword used before the broadcast of a request for assistance which is **urgent** but **not life threatening.**

This cannot be defined solely by the nature of the problem but only by considering the situation as a whole. The difference between a Mayday situation and one which would justify only a Pan Pan is the risk to the people involved. Engine failure could require a tow if the boat has no sails or if there is no wind but the crew may be in no danger while they wait. The skipper should expect to pay for this if life is not at risk. If the engine fails in poor weather and close to a lee shore then the situation could indeed be life threatening and a lifeboat might be required. Don't over-react by calling the rescue services too soon but don't wait until it is too late to complete a safe rescue if, for example, it begins to get dark or the weather deteriorates.

Each situation is unique. Consider the risks and all possible solutions before transmitting a Mayday, Pan Pan or routine call:

• How severe are the risks to the crew?

• Can you get yourself out of trouble?

• What sources of help are available?

• Are there other small boats in the area that could provide a tow or other assistance?

• Are you close to a marina that could send a boat to tow you in?

Any radio transmissions to a marina or another vessel should be made on the appropriate channel, naturally!

On a vessel fitted with VHF-DSC the **Pan Pan** on channel 16 could be preceded by a DSC **'Urgency Alert'** to all ships. The advantage of the Pan Pan over the routine call is that you will get immediate attention from the Coastguard even if they are busy. The call takes priority over all other calls except a Mayday. Another benefit is that on hearing the Pan Pan prefix or the 'Urgency Alert' all vessels in the area will be aware of your problem and so can offer to assist – or at least avoid hitting you!

The 'Urgency alert' sounds an alarm on all DSC sets within range which automatically retune to channel 16 for the voice message. A Pan Pan, unlike a Mayday, should be addressed **"all stations"** or **"all ships"**, but in the A1 area the Coastguard is most likely to reply.

Requests For Medical Advice

Vessels at sea can seek medical advice through the Coastguard. Call them on DSC or channel 16 as normal, or if the situation is urgent the call can be preceded by an **'Urgency Alert'** with the voice message prefixed **Pan Pan.** If the crew member is so ill or injured that you fear for their life then send a

DSC Urgency Alert to All Ships

❶ Select Call.

❷ Scroll to "All ships Urgency".

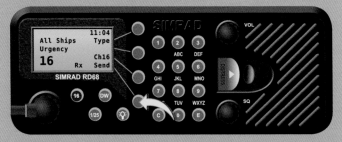

❸ Press send.

❹ Press E to confirm then transmit on channel 16, using high power.

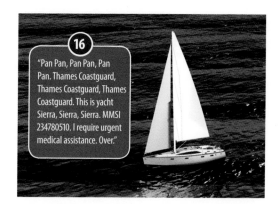

(16)

"Pan Pan, Pan Pan, Pan Pan. Thames Coastguard, Thames Coastguard, Thames Coastguard. This is yacht Sierra, Sierra, Sierra. MMSI 234780510. I require urgent medical assistance. Over."

Winching off

Sometimes used for "Medivac" following a Pan Pan.

distress alert followed by a **Mayday.** The Coastguard will organise a radio link to a doctor to give advice and then arrange a 'medivac' or medical evacuation for the casualty if requested by the doctor. Be prepared with information about the casualty and the position of the boat, destination and ETA.

Although the first transmissions use channel 16, subsequent communications may take place on another channel. You will not be required to choose this channel but must respond appropriately to the Coastguard's instruction. **Always acknowledge any change of channel before switching over.**

16 Securite ...'Safety Alert'

Messages containing safety information can be preceded by the proword **Securite** on channel 16, or a **'Safety Alert'** to all ships on DSC. The safety information will usually be read on a nominated channel, following the announcement on channel 16 or the DSC alert.

The most regularly heard of these broadcasts are made by the Coastguard stations. If it is just a routine repetition of the weather forecast then it would begin "all stations" but if a **new gale warning** is to be broadcast then the prefix Securite is used to draw attention to the importance of the information. Another example is a **navigation warning** or **small craft safety warning** about the detonation of an unexploded mine, not that uncommon in the Thames Estuary. In other regions information is broadcast by the Coastguard about naval firing exercises or submarine activity.

16

"All stations, All stations, All stations. This is Thames Coastguard, Thames Coastguard, Thames Coastguard. For a maritime safety information broadcast listen channel 86. Out."

16

"Securite, securite, securite. All stations, all stations, all stations. This is Thames Coastguard, Thames Coastguard, Thames Coastguard. For a new gale warning listen channel 86. Out."

Vessels at sea can use the same warning before an important safety message. A vessel engaged in underwater operations or survey work may broadcast a warning to other craft in the vicinity. A skipper having sighted a dangerous floating obstruction could do so too. A vessel with DSC could use a 'Safety Alert' to all ships which would sound an audible alarm on other DSC sets within range if justified in the circumstances .

Maritime Safety Information broadcasts by H.M. Coastguard

In 1999 the Coastguard took over the responsibility for the broadcast of all safety information and the facilitating of radio medical advice calls.

The MSI broadcasts are made at **three hourly** intervals, in the same way as the inshore water forecasts and gale warnings have been for several years. These forecasts have proved popular and accessible. The timings are certainly more 'user friendly' to the relaxed weekend sailor than the BBC shipping forecasts and often more relevant for coastal sailing. The different timings at adjacent Coastguard stations are to prevent mutual radio interference. The times at your local station may have changed from previous years because there has been a standardisation of times between local time and UT (GMT). All times are published in UT and should be correct in all new Almanacs. The local aerials allow the information to be received up to 30 to 40 miles out to sea as well as in local marinas and rivers.

The broadcasts are made on a nominated channel, following an announcement on channel 16. If there is a new gale or strong wind warning or one in force then the announcement is preceded by **Securite Securite Securite**. Weather forecasts are followed

DSC Safety Alert to all Ships

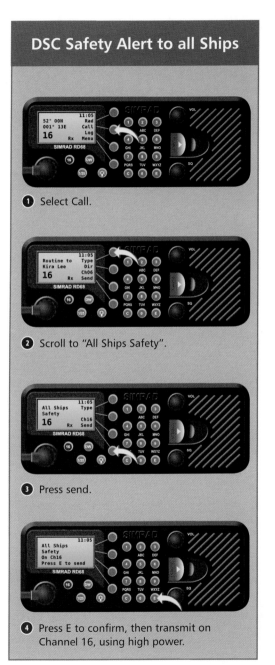

1 Select Call.

2 Scroll to "All Ships Safety".

3 Press send.

4 Press E to confirm, then transmit on Channel 16, using high power.

by the navigation warnings in force, any Negative Tide Surge Warnings and 'Gunfacts' and 'Subfacts' messages in areas of military or naval activity. The navigation warnings are all given numbers through the year and begin with the code letters WZ, so the messages are known as 'whiskey zulus'. The information given will be for the area of coast served by the Coastguard station and those areas adjacent to it. The system takes a little getting used to but is the most up-to-date information available and much, much quicker than waiting for the Notice to Mariners to be printed. Many are temporary warnings anyway. Those vessels with Navtex aboard will receive the WZ messages that way too.

Twice daily information from the BBC shipping forecast is included with the regular broadcast, usually the early morning and early evening broadcasts. Again the times are in the Almanac.

Urgent safety information is broadcast on receipt, preceded by a **Securite** and **DSC 'Safety Alert'** and then added to the routine broadcasts.

17 SIERRA ...the return!
An eventful homeward passage

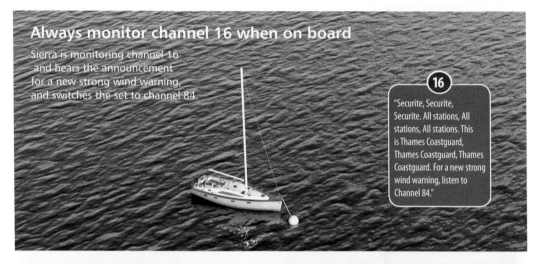

Always monitor channel 16 when on board

Sierra is monitoring channel 16
and hears the announcement
for a new strong wind warning,
and switches the set to channel 84.

16

"Securite, Securite, Securite. All stations, All stations, All stations. This is Thames Coastguard, Thames Coastguard, Thames Coastguard. For a new strong wind warning, listen to Channel 84."

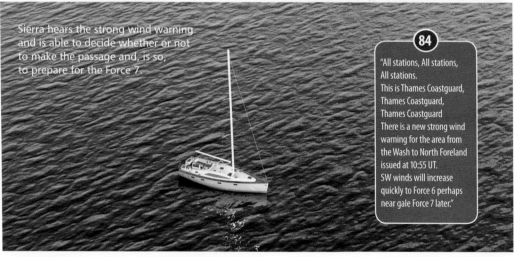

Sierra hears the strong wind warning
and is able to decide whether or not
to make the passage and, is so,
to prepare for the Force 7.

84

"All stations, All stations, All stations.
This is Thames Coastguard, Thames Coastguard, Thames Coastguard
There is a new strong wind warning for the area from the Wash to North Foreland issued at 10:55 UT.
SW winds will increase quickly to Force 6 perhaps near gale Force 7 later."

Sierra recieves a DSC distress alert

All vessels with VHF-DSC within range hear the DSC
distress alert. The Coastguard receives the alert too.

70

DSC DISTRESS ALERT.
Giving position, time, MMSI,
and nature of distress.

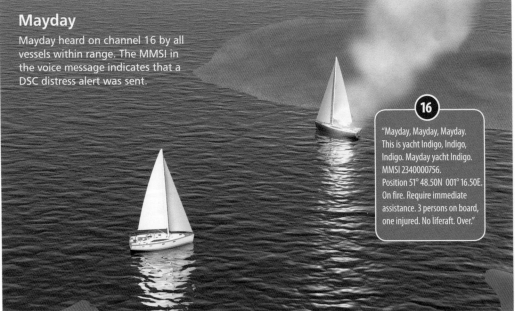

Mayday

Mayday heard on channel 16 by all
vessels within range. The MMSI in
the voice message indicates that a
DSC distress alert was sent.

16

"Mayday, Mayday, Mayday.
This is yacht Indigo, Indigo,
Indigo. Mayday yacht Indigo.
MMSI 2340000756.
Position 51° 48.50N 001° 16.50E.
On fire. Require immediate
assistance. 3 persons on board,
one injured. No liferaft. Over."

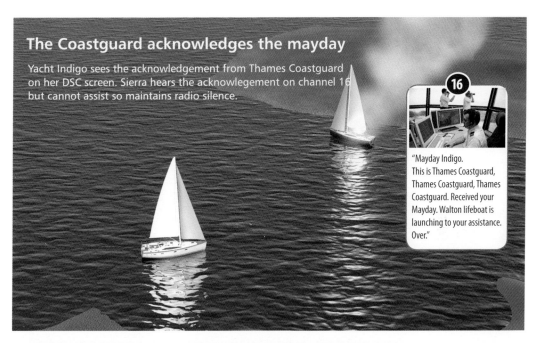

The Coastguard acknowledges the mayday

Yacht Indigo sees the acknowledgement from Thames Coastguard on her DSC screen. Sierra hears the acknowlegement on channel 16 but cannot assist so maintains radio silence.

16

"Mayday Indigo.
This is Thames Coastguard, Thames Coastguard, Thames Coastguard. Received your Mayday. Walton lifeboat is launching to your assistance. Over."

The lifeboat is launched

Sierra maintains radio silence.

16

"Mayday Indigo. Thames Coastguard. This is the Walton Lifeboat. ETA Indigo's position 10 minutes."

The lifeboat & helicopter are on scene

Sierra maintains radio silence.

16

"Mayday Indigo. Rescue 125. This is Thames Coastguard. Proceed to yacht Indigo and evacuate injured crew to hospital. Over."

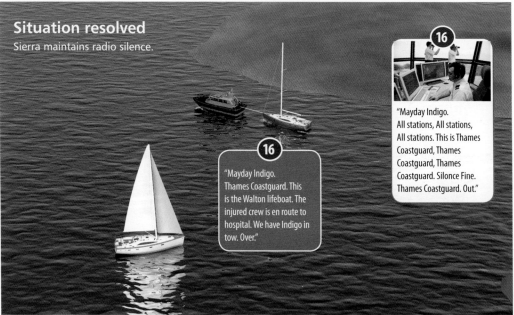

Situation resolved

Sierra maintains radio silence.

16

"Mayday Indigo. All stations, All stations, All stations. This is Thames Coastguard, Thames Coastguard, Thames Coastguard. Silonce Fine. Thames Coastguard. Out."

16

"Mayday Indigo. Thames Coastguard. This is the Walton lifeboat. The injured crew is en route to hospital. We have Indigo in tow. Over."

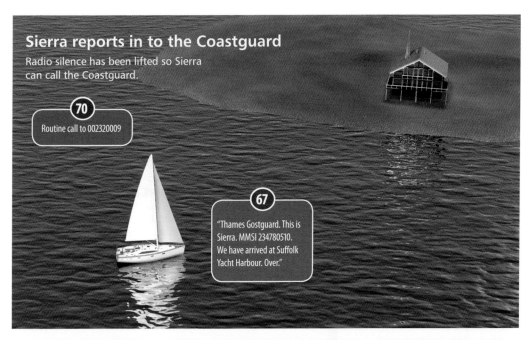

18 EPIRBs

An **E**mergency **P**osition **I**ndicating **R**adio **B**eacon is exactly what it says it is!

It is a portable, battery operated radio transmitter designed to initiate a distress alert on a pre-set frequency. It can be operated manually or activated automatically by a float-free device or hydrostatic release unit (HRU).

There are different types of EPIRB, with the cost and size varying a great deal. The variations are mainly to do with the frequency that is used for transmitting, the information included and how the beacon is deployed. The best can be expensive but have saved lives, as round-the-world yachtsman Tony Bullimore can testify.

The distress information transmitted from a **406 MHz** EPIRB is relayed by the COSPAS SARSAT satellites to a Rescue Co-ordination Centre and to **HM Coastguard**. Information on all alerts is initially sent to Falmouth Coastguard who then inform the Coastguard station nearest to the casualty's position. Another feature of the 406 MHz

beacon is the **additional information that can be transmitted**. This includes the identity of the vessel by means of a unique code and, in some cases, position from an inbuilt **GPS**. Information about an EPIRB carried on your vessel **must** be registered with the Coastguard and details should be sent to **HM Coastguard**, Falmouth. There is a form available from Coastguard stations for the registration of EPIRBs. The completed form should be sent to:

> The EPIRB Register
> HM Coastguard
> Pendennis Point
> Castle Drive
> Falmouth
> Cornwall
> TR114WZ
> Tel. 01326 211569

The EPIRB database at Falmouth holds details of the vessel so **do not loan** the EPIRB to another boat without this registration being changed. A good EPIRB is expensive but you can hire one for the period of an extended cruise. Check with your usual liferaft service and hire centre. In the case of a rented EPIRB the number will be registered to the company, who will know which beacon is on each vessel.

The EPIRB must also be listed on the ship's radio licence.

> If the beacon is **inadvertently activated** it must **not be switched off** until the nearest Coastguard station has been informed. This is to allow them to identify the signal correctly. To do this make a routine call using DSC, or if necessary channel 16.

19 SARTs

A **S**earch **A**nd **R**escue **T**ransponder, sometimes called a search and rescue **radar** transponder, is the part of the GMDSS which provides for the **locating** of vessels in distress or of their survival craft. It is complementary to an ERIRB, not an alternative.

When activated the SART will produce a series of up to 12 dots on a vessel's radar screen when it is interrogated by a 3 cm ship radar. These dots show the direction of the distress craft from the vessel operating the radar. The dots change to small arcs, and eventually circles, as the searching vessel or aircraft reduces the range to less than 1 mile. It is, in effect, an 'active' radar reflector in a similar way to a Racon. A SART is battery operated and can be switched on manually or automatically in combination with a float-free device such as a hydrostatic release unit (HRU). The life of the battery is about 96 hours on stand-by and 8 hours when transmitting. It operates on **9 GHz,** which means that the SART will be displayed on **all** radar sets, not just those on lifeboats and rescue helicopters. Report any sighting to the Coastguard.

SARTs are supplied with a pole to increase the height (and therefore the range and detectability) of the transponder. The minimum height recommended is 1 metre above sea level. **Do not use a radar reflector in close proximity to a SART** as the signal may be obscured. The reception is also affected by the height of the radar of the rescue craft, by the sea conditions and the skill of the radar operator.

Reception would typically be 5 miles for a ship and perhaps 30 – 40 miles for a searching aircraft, depending on height.

> The SART must be listed on the ship's radio licence.
> In the case of an **inadvertent activation:**
> • Switch off the transponder immediately.
> • Transmit a DSC 'Safety Alert' followed by an 'All Stations' call on channel 16, preceded by the word **Securite,** to cancel the false transmission. Include your MMSI in the voice message if DSC was used and your position.

16

"Securite, Securite, Securite.
All Stations. All Stations.
All Stations. is is yacht
Sierra, Sierra, Sierra. MMSI
234000756 SART operated
in error. In position... (give
latitude and longitude). Out."

20 NAVTEX

Another part of the GMDSS system is the **Navtex** receiver. This provides automatic reception of a world-wide English language **information service** covering **navigation, weather and other safety-related matters.**

For many years receivers operated only on 518kHz but dual frequency sets now also use 490kHz, allowing additional information to be included. The sets have a range of up to 300 miles, but a dedicated aerial is also required.

The sets fall into two broad groups: those where the information is displayed on a small screen where the user scrolls through to get the information, and older style ones which print out onto a roll of paper, rather like a fax machine. The information is in text: a Navtex receiver will not draw weather maps. The set is designed to be left switched on while the vessel is in use so that no information is missed.

The world-wide system is organised by dividing the world into **Navareas.** The UK falls into **Navarea 1.** Stations within that area broadcast at different times to avoid mutual interference and each of these stations is given an identification letter. This enables the user to select the most relevant station or stations and to change stations during a passage.

The messages that are broadcast are grouped by subject. These subject groups are identified by a letter code as shown in the table. The set can be programmed to exclude some of these messages. For example 'ice reports' or 'Loran messages. **Category A, B and D messages cannot be rejected by the receiver.**

Navtex is popular on vessels that frequently cruise long distances because of the system's range and the convenience of the information just 'arriving' without having to know the local times and frequencies. One drawback of the Navtex printer is that the amount of information can be considerable, so the machine will consume lots of paper.

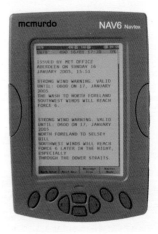

NAVTEX MESSAGE TYPES

TYPE	DESCRIPTION
A	Navigational warnings including: buoys out of position, lights unlit, new wrecks, floating debris, naval exercises (cannot be rejected by receiver)
B	Meteorological warnings (cannot be rejected by receiver)
C	Ice reports
D	Search And Rescue information, including Piracy and Armed Robbery warnings, (cannot be rejected by receiver)
E	Meteorological forecasts. Pattern of scheduled forecasts will vary from Navarea to Navarea
F	Pilot service messages
G	DECCA messages
H	LORAN messages
I	OMEGA messages
I	Satnav messages
K	Other electronic Navaid messages (messages concerning radio navigation services)
L	Navigational warnings for mobile drilling rig movements (should not be rejected by receiver)
V	Amplifying navigational warning information initially announced under message type A
W	Special services – trial allocation
X	Special services – trial allocation
Y	Special services – trial allocation
Z	No messages on hand. May be broadcast when applicable to confirm correct receiver operation

21 Radio channels

INTERNATIONAL VHF FREQUENCIES

Channel Number	Transmitting Frequency MHz		Intership	Port Operations and Ship Movement		Public Correspondence
	Ship Stations	Coast Stations		Single Frequency	Two Frequency	
60	156.025	160.625			X	X
01	156.050	160.650			X	X
61	156.075	160.675			X	X
02	156.100	160.700			X	X
62	156.125	160.725			X	X
03	156.150	160.750			X	X
63	156.175	160.775			X	X
04	156.200	160.800			X	X
64	156.225	160.825			X	X
05	156.250	160.850			X	X
65	156.275	160.875			X	X
06	156.300		X			
66	156.325	160.925			X	X
07	156.350	160.950			X	X
67	156.375	156.375	X	X	Small Ship	Safety Channel
08	156.400		X			
68	156.425	156.425		X		
09	156.450	156.450	X	X		
69	156.475	156.475	X	X		
10	156.500	156.500	X	X		
70	156.525	156.525	Digital	Selective	Calling	Only
11	156.550	156.550		X		
71	156.575	156.575		X		
12	156.600	156.600		X		
72	156.625		X			
13	156.650	156.650	X	X		
73	156.675	156.675	X	X		
14	156.700	156.700		X		
74	156.725	156.725		X		

Channel Number		Transmitting Frequency MHz		Intership	Port Operations and Ship Movement		Public Correspondence
		Ship Stations	Coast Stations		Single Frequency	Two Frequency	
15		156.750	156.750	X	X		
	75	156.775			X		
16		156.800	156.800	Distress	Safety	and	Calling
	76	156.825			X		
17		156.850	156.850	X	X		
	77	156.875		X			
18		156.900	161.500		X	X	X
	78	156.925	161.525			X	X
19		156.950	161.550			X	X
	79	156.975	161.575			X	X
20		157.000	161.600			X	X
	80	157.025	161.625			X	X
21		157.050	161.650			X	X
	81	157.075	161.675			X	X
22		157.100	161.700			X	X
	82	157.125	161.725		X	X	X
23		157.150	161.750			X	X
	83	157.175	161.775		X	X	X
24		157.200	161.800			X	X
	84	157.225	161.825		X	X	X
25		157.250	161.850			X	X
	85	157.275	161.875		X	X	X
26		157.300	161.900			X	X
	86	157.325	161.925		X	X	X
27		157.350	161.950			X	X
	87	157.375			X		
28		157.400	162.000			X	X
	88	157.425			X		
AIS 1		161.975	161.975				
AIS 2		162.025	162.025				

NOTES
• For intership use 06, 08, 72 and 77.
• Channel 70 must NEVER be used for voice communication.

22 Glossary

ACKNOWLEDGEMENT: The reply to a Mayday. It implies that assistance will be provided either practically, as by a vessel at sea, or via a third party when the Coastguard organises a lifeboat or helicopter.

AMERC: Association of Marine Electronic and Radio Colleges. They run courses for professional radio operators and for the Long Range Certificate.

BROADCAST: To transmit on the radio without expecting a reply. The BBC broadcasts both speech and music, as did the pirate pop radio stations in the North Sea. The Coastguard broadcast Maritime Safety Information every four hours. It is prohibited for VHF radio users to make broadcasts in this way.

CEPT: Conference of European Posts and Telecommunications designed the syllabus for the SRC.

COAST RADIO STATIONS: These stations were run by British Telecom to provide a phone link between the VHF and the subscriber ashore. Now largely ceased.

CONVENTION SHIP: These are the vessels which have to fit all the equipment required under GMDSS for the area of the world in which they operate. They include ships over 300 GRT (gross registered tons) and passenger vessels which carry more than 13people. They are also known as 'compulsory fit' vessels. All other vessels, including all private boats, are 'voluntary fit' vessels and may chose to fit some or all of the recommended equipment.

COSPAS/SARSAT: Search and rescue satellite organisations. These receive the signals from ERIRBS and relay the information to a rescue centre.

DSC: Digital selective calling

DUPLEX: Method of radio working that requires two aerials and two frequencies. It was used by Coast Radio Stations but rarely on yachts or motor cruisers

DUAL WATCH: A facility on the radio to allow the monitoring of channel 16 and another nominated channel.

EPIRB: Emergency Position Indicating Radio Beacon

GMDSS: Global Maritime Distress and Safety System

GPS: Global Position System. If this is interfaced with a DSC controller then the position and time can be included automatically in the distress alert.

HRU: Hydrostatic Release Unit. A device attached to a liferaft or EPIRB to provide a weak link to ensure that the liferaft or EPIRB will automatically float free in the event of the vessel sinking.

IMO: The International Maritime Organisation is the world organisation which developed GMDSS.

ITU: International Telecommunications Union.

INMARSAT: The International Marine Satellite Organisation operates satellites for communications for ocean going vessels. Some round the world yachts use Inmarsat 'C' This is capable of sending text messages only.

LUT: Local User Terminal. The reception station for satellite signals relaying information from EPIRBs.

MCA: Maritime and Coastguard Agency

MF: Medium frequency radio. A radio capable of transmitting over longer range then a VHF radio and requiring a different certificate and licence. Advisable for passages outside the AI area.

MHz: Megahertz

MSI: Marine Safety Information broadcast. Regular broadcast made by the Coastguard which includes weather forecasts and navigation warnings.

MMSI: Maritime Mobile Service Identity. The 9 digit number programmed into the set.

MRSC: Maritime Rescue Sub-Centre. A Coastguard station manned 24 hours a day, 365 days a year. These stations are grouped, with the regional management being based at an MRCC (Maritime Rescue Co-ordination Centre).

NAVTEX: Radio that receives text messages about maritime safety.

NMEA: The maritime industry standard for interfacing of electronic equipment.

Ofcom: Office of Communications. They are the radio licensing authority.

PLB: Personal Locator Beacon. A type of mini-EPIRB worn by an individual crew member, ideally with a locating device on board the vessel. Used on round-the-world ocean-racing vessels.

RCC: Rescue Co-ordination Centre. Run by the military for the co-ordination of non-civilian search and rescue. Information from the LUT is sent to the RCC and then passed to the Coastguard, if it does not involve the military.

ROUTINE: Communications not concerning distress or a request for assistance.

RYA: Royal Yachting Association. The RYA issues Short Range Certificates and qualifies instructors to teach and assess these radio courses. The courses are run largely by sailing schools and the centres that run RYA shorebased courses.

SART: Search and Rescue radar Transponder.

SIMPLEX: The method of radio working most common on yachts and motor cruisers. It uses the same aerial for transmitting and receiving, switching from one to the other when the PTT switch on the microphone is pressed.

SQUELCH: The control on the radio that reduces background radio signals which cause interference.

SRC: Short Range (radio) Certificate. The certificate which is replacing the simple VHF certificate.

PTT: The Press To Talk switch on the microphone which changes the aerial from receive to transmit. It must be released to receive the reply.

VHF-DSC: A radio telephone with a DSC controller either incorporated into the set or as an additional set.

VHF: Very High Frequency radiotelephone. Suitable for communications between boats up to 15 miles and between boats and shore stations over ranges of up to 40 miles, depending on the aerial height of both the transmitting and receiving stations.

VTS: Vessel Traffic Service. Name used by some harbour authorities.

Questions:

1. What voice message should follow a distress alert sent on DSC?

2. Which of the following would justify a distress alert and Mayday?

a) A crew member with a broken arm, who is in a lot of pain.

b) Seeing a red flare or orange smoke signal.

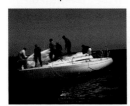 

c) A broken down yacht unable to sail into the marina.

d) A broken down motor cruiser dragging its anchor towards rocks in a gale.

3. The yacht Blackbird has hit a sandbank in position 51° 45.50N 001° 58 .70E. It is pounding heavily and in danger of breaking up. On board are the skipper and two crew, who are preparing the liferaft. The MMSI is 234007839. Write the Mayday call and message that you would send after the distress alert has been sent.

4. The voice Mayday should be sent on:
   a) channel 70
   b) channel 16
   c) channel 67
   d) channel 80

5. What is the meaning of:
   a) Seelonce Mayday
   b) Seelonce Finee

6. The motor cruiser Swift has accidentally sent a distress alert:
   a) What action should they take?
   b) Write the voice message that they should send.

7. For which of the following would it be incorrect to send a distress alert and Mayday?
   a) A crew member has been hit on the head and is drifting into unconsciousness.
   b) A boat has seen a semi-submerged object that might be a hazard to other vessels.
   c) A crew member has fallen overboard and cannot be rescued immediately.
   d) A fire in the engine room of the boat.

8. The yacht Sierra receives a distress alert and hears the Mayday from the yacht Indigo followed by the acknowledgement from Thames Coastguard. What VHF message, if any, should they send?
   a) They cannot assist.
   b) They can assist.

9. A distress alert will be re-transmitted about every 4 minutes, until it is digitally acknowledged by a class A or B DSC controller.
   True or false

10. **A yacht or motor cruiser could do this with their class D DSC set.**

    True or false.

11. **In what circumstances should a Mayday Relay be sent?**

12. **Which DSC Alert includes the position of the vessel, if the radio is linked to a GPS?**
    a) A Safety Alert.
    b) An Urgency Alert.
    c) A Distress Alert.
    d) All of them.

13. **An Urgency Alert could be sent before transmitting a Mayday Relay, if necessary.**

    True or false

14. **Why would sending a distress alert before transmitting a Mayday Relay be seriously confusing for the Coastguard?**

15. **What voice message would normally follow an Urgency Alert?**
    a) Any message that seems important
    b) A securite
    c) A pan pan
    d) A mayday

16. **A Pan Pan message could be used if**
    a) A vessel has broken down and requires some assistance, the situation is urgent, but not life threatening.
    b) A vessel and its crew are in grave and imminent danger because of a fire onboard.
    c) A yacht has engine failure, but can sail onto its mooring.
    d) A motor cruiser has lost one engine, had to slow down, so will be late arriving.

17. **A crewmember on the motor cruiser Swift is experiencing severe stomach pains.**
    a) Who could the skipper contact for medical advice?
    b) Write the radio call the skipper should make if the pain seemed to seriously deteriorate.

18. **You receive a Safety Alert while your DSC radio is set to channel 80 because you are about to call a marina. The radio will retune to channel 16.**

    True or false

19. **Regular Maritime Safety Information broadcasts are made by the Coastguard.**
    a) What information is included?
    b) When are MSI broadcasts preceded by a Safety Alert using DSC?

20. **Low power is use in routine calls to**
    a) limit the range of the transmission
    b) or save battery power

21. **It is OK to use the VHF to make arrangements to meet in the pub.**

    True or false

22. **Which of the following does not need to be listed on the Ships Radio Licence**
    a) VHF
    b) Radar
    c) portable VHF
    d) Navtex
    e) EPIRB
    f) SART

23. **Since October 1st 2006 a Ship Radio Licence is no longer required.**

    True or false

24. **In the UK which channel should be used to call a marina?**
    a) Channel 80
    b) Channel 16
    c) Cannel 67
    d) Channel 70

25. **It is legal to use the portable VHF in the dinghy under the Ships Radio Licence, but not ashore.**

    True or false

26. It's correct to say "over and out" at the end of radio working, as they do on TV programmes.

    True or false

27. An EPIRB must be registered with the Coastguard and mentioned on the Ships Radio Licence.

    True or false

28. If an EPIRB is activated all vessels in the area will know that a vessel is in distress immediately.

    True or false

29. What action should be taken if an EPIRB is activated accidentally?

30. A 406MHz EPIRB sends signals to satellites and is a world-wide system of distress alerting which is highly recommended for passages into the A2 area.

    True or false

31. The MMSI must be programmed into radio to make the DSC work correctly. How is the number obtained?

32. You want to call a boat but do not know the MMSI. Should you
    a) Call by voice on channel 70?
    b) Do a DSC routine call, but without their MMSI?
    c) Call on channel 16 and suggest an intership channel?
    d) Call on an intership channel, and hope the other boat will hear?

33. What are the recommended intership channels?

34. You have called the Coastguard on a routine DSC call which they have acknowledged. You should now call them by voice. What information should be included in addition to the boat's name in the first call?

35. A SART will produce a target on all radar sets, not just those of lifeboats and SAR helicopters.

    True or false

36. If s SART is operated in error should you make a securete broadcast to all stations on channel 16 or call the Coastguard?

37. What is channel 13 used for?

38. An MMSI starting with 00 is a...

    Complete this sentence.

39. A navtex receiver will draw a weather map.

    True or false

40. You are approaching a port where the regulations state that you should call the port operations centre. You do not know the channel to use. Should you
    a) Look it up in the almanac?
    b) Use channel 16?
    c) Use DSC?
    d) Not call at all and hope that you don't get run down by a ship or prosecuted!

41. To operate a VHF-DSC someone on the boat must have an SRC. The courses and assessments are organise by
    a) The RYA?
    b) Ofcom?
    c) The Coastguard?
    d) BT?

42. The range of a VHF set is affected by
    a) The height of the aerial?
    b) The transmitting power of the radio?
    c) Obstructions to the signal, such as the horizon or land?
    d) All of these?

43. Two yachts with an average aerial height, would expect to be able to communicate over
    a) 5 to 10 miles?
    b) 15 to 20 miles?
    c) 20 to 30 miles?
    d) 30 to 40 miles?

44. The dual watch, tri-watch and scan facility allow more than one radio channel to be monitored. Before transmitting it is important to switch this off and select the correct channel.
    a) True or false
    b) Is it permitted to call another boat on the intership channel?

45. Spell the boat name INDIGO using the phonetic alphabet.

46. You call a marina on channel 80. The operator asks you to "stand by". What should you do?

47. Should you do a radio check every day with the Coastguard?

48. The preferred method of calling for the Coastguard is:
    a) Channel 16?
    b) Channel 67?
    c) DSC?
    d) Mobile phone?

49. You need to call the motor cruiser Swift, who does not have DSC. You have tried their mobile phone with no success so you have to use channel 16. Write your initial call.

50. A motor cruiser with a vertically mounted aerial at 3m should be able to hear the weather forecast read by the Coastguard at 30-40 miles, the same as yacht.

    True or false

51. If you can answer all these questions you have read the book very carefully, should pass the SRC assessment easily and be a good and considerate user of the VHF-DSC radio.

    True or false

Answers

1. A voice Mayday should follow the Distress Alert.

2. d) <u>Your</u> situation must be life threatening for a Mayday to be sent.

3. Mayday, Mayday, Mayday
    This is yacht Blackbird, Blackbird, Blackbird
    Mayday yacht Blackbird. MMSI 234007839.
    In position 51° 45.50N 001° 58.70E.
    Aground and breaking up.
    Require immediate assistance.
    3 persons on board.
    Abandoning to liferaft.
    Over.

4. b) Send a Mayday on channel 16.

5. a) Seelonce Mayday may be said by the controlling station, usually the Coastguard, or the casualty vessel to impose radio silence if it is necessary.
    b) Seelonce finee may be said by the controlling station to lift radio silence once the situation is resolved.

6. a) Turn the set off, to cancel the automatic repetitions, and then back on again.

    b) On channel 16
    All stations, all stations, all stations
    This is Swift, Swift, Swift
    MMSI 233077310.
    Cancel my distress alert.
    Distress alert sent in error.
    Out.

7. b) A hazard that has been sighted is not a distress situation. Report it to the Coastguard, or the skipper could do a securite broadcast if the danger to other craft was immediate and the Coastguard cannot be contracted.

8. a) Maintain radio silence on channel 16 or any other channel being using for the rescue. Using other channels, such as calling a marina on channel 80 is allowed.

    b) Mayday Indigo
    Thames Coastguard
    This is yacht Sierra, Sierra, Sierra.
    Proceeding to assist. ETA 10 minutes.
    Over.

9. True

10. False

11. A Mayday Relay is sent by a vessel, not itself in distress, to pass on distress information. It is used if the vessel in distress is unable to send a Mayday or if the Mayday is not acknowledged.

12. c) A distress alert is the only DSC alert that includes the position.

13. True

14. Sending a distress alert before a Mayday relay would be misleading for the Coastguard because the MMSI and position of the sender would be transmitted, not that of the casualty vessel.

15. c) An Urgency Alert should be followed by a Pan Pan message.

16. a) Pan Pan means that assistance is required, it is urgent, but not life threatening.

17. a) The Coastguard can arrange for a doctor to give medical advice in the case of sudden illness or injury on a boat.

    b) Pan Pan, Pan Pan, Pan Pan
    Thames Coastguard, Thames Coastguard, Thames Coastguard,
    This is motor cruiser Swift, Swift, Swift
    I require urgent medical assistance. Over.

18. True. Receiving any all stations DSC alert; distress alert, Urgency Alert or Safety Alert switches the receiving radio to channel 16.

19. a) Coastguard MSI broadcasts may include: gale warnings, strong wind warnings, local and regional weather forecasts and navigation warnings.

    b) If there is a new gale warning the MSI broadcast may be proceeded by a Safety Alert.

20. a) Use low power to limit the range of routine transmissions.

21. False. The VHF should not be used to make social arrangements.

22. d) Navtex. The licence covers units that transmit, Navtex is a radio receiver.

23. False. A licence is still required, but since October 2006 it has been free if arranged over the internet. Not having a licence could result in prosecution.

24. a) Use channel 80 to call marinas in the UK.

25. True. The boat's portable radio can be used in the dinghy, but not ashore.

26. False. 'Over and out' makes no sense. 'Over' means please reply and 'out' means goodbye! The TV programmes are wrong.

27. True. EPIRBs should be listed on the licence, and registered with the Coastguard for Search and Rescue purposes.

28. False. A 406 Mhz EPIRB sends a signal via satellites. The information is passed to the Coastguard. Vessels in the area will be unaware that it has been activated.

29. If an EPIRB is activated in error contact the Coastguard as soon as possible. Switch it off only when told to do so.

30. True. EPIRBs are highly recommended for offshore passages.

31. The MMSI is allocated by Ofcom and can be requested with the ship radio licence or at a later date.

32. c) Without the MMSI, or if the boat to be called does not have DSC, the only option is to call on channel 16 and suggest an intership channel. Even better would be to agree the intership channel in advance and both boats monitor that using the dual watch.

33. 6, 8, 72, and 77 are the recommended intership channels.

34. The MMSI.

    The first voice call on channel 67 to the Coastguard, following the DSC call and their acknowledgement, should include the MMSI. This is because they may receive several DSC calls within a few minutes and the MMSI is required to identify which one is speaking. This is only necessary initially, and not necessary at all if making a routine call to another boat.

35. True. A SART will be displayed by all radar sets.

36. Make a securite broadcast to all stations, giving your position, so all vessels who may have seen it on their radar are aware.

37. Channel 13 is used for bridge to bridge communication on matters of navigational safety on ships.

38. An MMSI starting with 00 is a coast station, such as a Coastguard.

39. False. A Navtex provides text information only.

40. a) Check the port operations channel in the almanac.

41. a) The RYA organises the SRC courses through RYA centres.

42. d) The height of the aerial, the transmitting power and obstructions all affect the range of the set.

43. b) 15 - 20 miles.

44. a) True. Switch off dual or tri-watch before transmitting.

    b) Yes. Calling on a prearranged intership channel is permitted and a good idea, allowing vessels without DSC to avoid routine calling on channel 16.

45. Indigo, I spell:

    India November Delta India Golf Oscar.

46. If asked by the station you call to 'stand by', wait on the same channel until they call you back.

47. No. Doing a radio check every day is probably unnecessary, and certainly should not be done with the Coastguard. Calling a marina or another boat, avoiding channel 16, causes less interference.

48. c) The preferred method of calling for the Coastguard is DSC, or channel 67 in the Solent. Only use channel 16 for routine calling if there is no alternative.

49. Swift, Swift
    This is Solo
    Channel 72
    Over.

50. False. The range will be less because of the lower aerial.

51. Very, very true. Congratulations!

Sara

23 Useful Addresses

For information about training:

Royal Yachting Association
0845 345 0377
E mail certification@rya.org.uk
www.rya.org.uk

To apply for a radio licence contact:

www. radiolicencecentre.co.uk
or www.ofcom.org.uk
020 7981 3131

For the registration of EPIRB:

The EPIRB Registry
HM Coastguard, Falmouth
epirb@mcga.gov.uk
01326 211569

Mayday Procedure Card

DSC distress alert

Open the red cover

Press the red button

If time select the relevant distress situation,

Press and hold the red button through the countdown

Wait no more than 15 seconds then send the voice message on channel 16, high power.

Mayday call

Mayday Mayday Mayday

This is_____ _____ _____ (give name three times)

Mayday message

Mayday (Boat's name) _____ (give once)

MMSI_____

In position_____

(Give latitude and longitude or range and bearing from a named object)

Nature of distress

Require immediate assistance

___ persons on board

Other VITAL information (abandoning vessel, no liferaft etc)

Over

Lifeboats

'Flat calm or force 10. I always wear one.'

Whether they're training or out on a shout, RNLI crew members always wear lifejackets. It's a rule informed by years of experience. They know that, whatever the weather, the sea's extremely unpredictable – and can turn at a moment's notice. They see people caught out all the time. People who've risked, or even lost their lives as a result. The fact is, a lifejacket will buy you vital time in the water – and could even save your life. But only if you're wearing it.

For advice on choosing a lifejacket and how to wear it correctly, call us on 0800 328 0600 (UK) or 1800 789 589 (RoI) or visit our website rnli.org.uk/seasafety/lifejackets

Useless unless worn